VICTORIA

Physical Environment and Development

VICTORIA
Physical Environment and Development

edited by

Harold D. Foster

Western Geographical Series, Volume 12

Department of Geography
University of Victoria
Victoria, British Columbia
Canada

1976 University of Victoria

Western Geographical Series, Volume 12

EDITORIAL ADDRESS
Harold D. Foster, Ph.D.
Department of Geography
University of Victoria
Victoria, British Columbia
Canada

Publication of the Western Geographical Series has been generously supported by the Leon and Thea Koerner Foundation, the Social Science Research Council of Canada, the National Center for Atmospheric Research, the International Geographical Union Congress, the University of Victoria and the National Research Council of Canada.

ISSN 0315—2022
ISBN 0—919838—02—2

ACKNOWLEDGMENTS

The editor would like to acknowledge that this book has been published with the help of a grant from the National Research Council of Canada.

Many people have contributed towards its successful publication. The layout and design of the cover and contents, together with the cartographic and photographic work was undertaken by Mr. Ian Norie, assisted by Mr. Ole Heggen and Mr. Ken Quan. Mrs. Alison Griffith also provided valuable assistance. The manuscript was typed by Mrs. Laurel Carr and Miss Louise A. Evans. Their kind cooperation is gratefully acknowledged.

Harold D. Foster

University of Victoria

Victoria, B.C.

May, 1976

PREFACE

Centrifugal forces have long threatened geography with disin-
tegration. For over a century, physical geography has been dominated
by a desire to study natural landforms, vegetation and climate. Man,
where he interfered with such pristine elements, was considered an unfortunate irritant. Under these circumstances, the development of a
widening gulf between physical and social geography was inevitiable.

Fortunately, the growing demand for relevancy is leading to a
greater unity of purpose in geography. The nine chapters which follow,
although written by authors predominantly interested in the physical
milieu, are essentially applied, exploring the interface between physical
and social systems in the Greater Victoria area. Although limited to
an examination of this unique region, Victoria: Physical Environment and
Development presents a philosophy, butressed by several applied models,
which has far wider relevance and should be of interest to physical and
social scientist alike.

Harold D. Foster

University of Victoria
Victoria, B.C.
May, 1976

PLATE 1
Aerial Mosaic of the Saanich Peninsula.

TABLE OF CONTENTS

LIST OF TABLES

LIST OF FIGURES

xiii

LIST OF PLATES

PLATE 2
Victoria's Inner Harbour and
Central Business District.

CHAPTER 1

THE PHYSICAL GEOGRAPHY OF VICTORIA,
CIRCA 1860,
AS PERCEIVED BY COLONISTS, MAPMAKERS AND VISTORS

Charles N. Forward

University of Victoria

INTRODUCTI'ON

The physical geography of the Victoria area was described in some
detail during the first two decades of European settlement. Numerous
reports, essays and books were written that devoted considerable attention
to the physical geography, and a number of maps were produced. A century and a quarter of European occupance brought about far-reaching
changes in the landscape. Certain physical elements were affected far
more than others, particularly vegetation, soils and drainage within the
built-up areas of metropolitan Victoria (Figure 1,1). Climate, geological
structure and major topographic features remain essentially unaltered by
man, though urban development, mining, quarrying and road construction
have caused minor, localized changes, even in these enduring elements.
Whether or not landscape elements were altered, there was a significant
change in the perception of physical features. The present perception of
the physical environment is based on lengthy scientific observations and
records, as well as the work of many researchers who had the benefit of
greater knowledge of physical processes than the observers of a century
ago.

After James Douglas established Fort Victoria in 1843 settlement
proceeded very slowly until the gold rush period beginning in 1858. As
a result, relatively little was written about the Colony of Vancouver Island until it became famous as the point of entry to the British Columbian

1

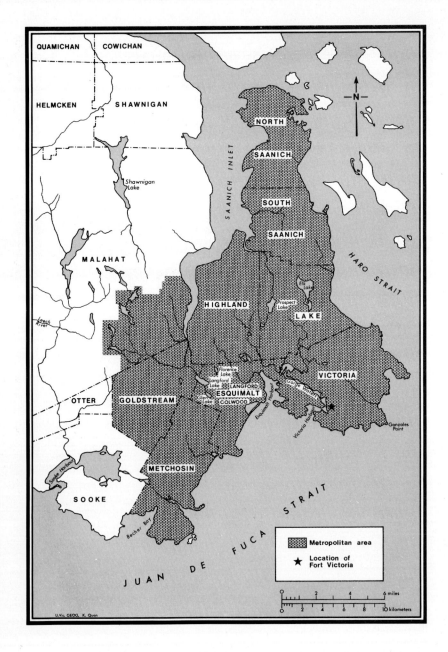

FIGURE 1,1   Land districts of the Victoria area.

2

gold fields and, hence, newsworthy in the eyes of the publishers. A flurry of books and articles about Vancouver Island and British Columbia appeared in the late 1850's and early 1860's. Detailed maps of the land districts were printed and the stage was set for the expected influx of new settlers. The perception of the physical geography of the Victoria area around 1860 had a great impact on the attitude of the Mother Country towards her eastern Pacific island colony and a continuing influence on the self-image, capabilities and ambitions of the colony itself.

The twofold purpose of this study is to determine as accurately as possible the physical geography of the Victoria area before its profound alteration by urban development and to portray the physical geography as perceived by the writers and mapmakers of more than a century ago. Ancillary to these basic endeavours is the attempt to compare the land-scape of 1860 with that of the present and to interpret the effects of the landscape perceptions enunciated at that time on the future development of the colony.

GENERAL CHARACTER OF THE AREA

Many years of Indian occupance left little imprint on the land-scape, as one would expect in view of the fact that, historically, they were neither cultivators nor pastoralists. Their fishing, hunting and gathering activities were their livelihood and their demands upon the luxuriant forests were relatively modest — cedar timbers and planks for house construction, cedar logs for dugout canoes and maples for making paddles, tool handles and dishes. Small cleared areas sufficed for their villages, generally situated along the coast. It has been suggested that the large tract of open land around Victoria Harbour that so influenced James Douglas' choice of site for the post, "resulted from the constant digging by Indian women with their sharpened sticks for the roots of the camass".[1] The camass is a deep-rooted, flowering bulb native to the

3

region, the bulbs of which were a favorite food of the Indians. As the
camass grew in great profusion around the harbour, the Indian name for
this district was "Camosun", meaning "place for gathering camass".[2]
While this may have been a contributing factor locally in causing the
open terrain, it was the nature of the Garry oak-grassland vegetation
to exhibit a parkland character. Close to their villages the Indians may
have cultivated small plots of potatoes before Fort Victoria was established,
having learned the technique through earlier contacts with fur traders.
The first independent colonist in the area, Colonel W. Colquhoun Grant,
reported in 1857 that, "the potato is almost universally cultivated by all
the savage tribes on the south of Vancouver Island as well as on the op-
posite mainland," and he went on the say that they had been doing so for
a long time.[3]

    The establishment of Fort Victoria marked the entry of European
settlement under the guidance and control of the Hudson's Bay Company.
Anticipating a boundary settlement along the 49th parallel between the
United States and British territories, the Company wished to establish a
new headquarters depot for its fur trading activities on British soil, eventually to replace Fort Vancouver near the mouth of the Columbia River.
In the summer of 1842 James Douglas was commissioned with the task of
finding the most suitable location for the new fort on the southern end
of Vancouver Island. He commented on the physical geography of various
sites inspected closely in his report to Chief Factor John McLoughlin.

    Of Sooke Harbour he spoke of:

> . . . the shores being high, steep, rocky, and
> everywhere covered with woods. In ranging
> through the forest we found one small plain con-
> taining 3 or 400 acres of land at the distance of
> one mile from the harbour; but the rest of the
> country in its neighbourhood appeared to consist
> either of wood land or rocky hills.[4]

4

Visiting Metchosin, one and one-half miles to the east, he stated, "the extent of clear ground is much too small for the demands of a large establishment, and a great part of what is clear is poor stony land with rolling surface, so that on the whole it would not do for us."[5] He considered Esquimalt as one of the best harbours on the coast, but in other respects as possessing little attraction:

> Its appearance is strikingly unprepossessing, the outline of the country exhibiting a confused assemblage of rock and wood.
> More distant appear isolated ridges thinly covered with scattered trees and masses of bare rock, and the view is closed by a range of low mountains that traverse the island at the distance of about 12 miles. The shores of the harbour are rugged and precipitous, and I did not see one level spot, clear of trees, of sufficient extent to build a large Fort upon. There is in fact no clear land within a quarter of a mile of the harbour, and that lies in small patches here and there, on the aclivites and bottoms of the rising ground. At a greater distance are two elevated plains, on different sides of the harbour, containing several bottoms of rich land, the largest of which does not exceed 50 acres, of clear space, much broken by masses of limestone and granite.[6]

By far the most suitable site for the fort, Douglas concluded, was at Victoria Harbour:

> . . . at Camosack there is a pleasant and convenient site for the establishment, within 50 yards of the anchorage, on the border of a large tract of clear land, which extends eastward to Point Gonzalo at the south-east extremity of the Island and about 6 miles interiorly, being the most picturesque and decidedly the most valuable part of the Island that we had the good fortune to discover. . .
> More than two-thirds of this section consists of prairie land and may be converted either to the purposes of tillage or pasture, for which I have seen no part of the Indian Country better adapted; the

5

rest of it with the exception of the ponds of water
is covered with valuable oak and pine timber. I
observed, generally speaking, but two marked
varieties of soil on these prairies, that of the best
land is a dark vegetable mould, varying from 9 to
14 inches in depth, overlaying a substrate of greyish
clayey loam which produces the rankest growth of
native plants that I have seen in America. The other
variety is of inferior value, and to judge from the less
vigorous appearance of the vegetation upon it, naturally more unproductive.

   Both kinds however produce abundance of grass,
and several varieties of red clover grow on the rich
moist bottoms.

   In two places particularly we saw several acres
of clover, growing with a luxuriance and compact-
ness more resembling the close sward of a well
managed lea than the produce of an uncultivated
waste.

   Being pretty well assured of the capabilites of the
soil as respects the purposes of agriculture, the climate being also mild and pleasant we ought to be
able to grow every kind of grain raised in England.[7]

He was greatly impressed with Gorge Waters and the trees along its shores:

   The canal of "Camosack" is nearly six miles long,
and its banks are well wooded throughout its whole
length, so that it will supply the establishment with
wood for many years to come, which can be conveyed
in large rafts, with very little trouble, from one extreme of the Canal to the other.[8]

Douglas went on to report that there was a natural supply of fresh water,
a small stream at a distance of 300 paces from the proposed location of
the fort and a lake at 800 yards away.[9] It is evident that the oak-grassland association was particularly widespread on the Victoria peninsula,
offering far less impediment to the clearing of land for agricultural use
than the heavily forested land so common on southern Vancouver Island.
The fort was established near the shore of Victoria Harbour and within

6

twenty years the town that had sprung up outside the fort counted a population of several thousand people and was incorporated as a city (Figure
2,1). By that time, also, agricultural settlement had penetrated most of
the areas of fertile land within a radius of twenty-five miles of the fort.
But the amount of land actually cleared beyond the confines of the town
itself was relatively insignificant, leaving the landscape essentially in
its wilderness condition.

Vancouver Island in general was considered unfavourable by many,
Victoria being looked upon as a virtual oasis of attractive elements placed
against a bleak background. Based on his experience on Vancouver Island in the early 1850's, Colonel Grant proclaimed:

> The general aspect of the country throughout the
> island from the seaward is particularly uninviting.
> Dark frowning cliffs sternly repel the foaming sea,
> as it rushes impetuously against them, and beyond
> these, with scarcely any interval of level land,
> rounded hills, densely covered with fir, rise one
> above the other in dull uninteresting monotony;
> over these again appear bare mountains of trap
> rock, with peaks jagged like the edge of a saw, a
> veritable Monserrat, forming a culminating ridge,
> which may be said to run with little intermission,
> like a back bone, all down the centre of the island,
> from the northern to the southern extremity; nor
> does a nearer approach present one with many more
> favourable features in the aspect of the country....[10]
> The hills are steep and rugged; the valleys narrow
> and shallow; the rocks are sometimes bare, sometimes
> covered with a scant growth of timber: but in no case,
> that I have seen, does the surface of the interior of
> the island, either in its nature or its position, admit
> of being applied to any more useful purpose than to
> furnish matter for the explorations of a geologist.
> From these regions, which are wild without being
> romantic, and which, from the absence of any bold
> outline, never approach to the sublime or the beau-
> tiful, the traveller loves to descend to the smiling
> tracts which are occasionally to be met with on the
> sea-coast. In one of these Victoria is situated, and

FIGURE 2,1
The northeastern bastion of
Fort Victoria circa 1860.

> it is from a visit to it, and its neighbourhood,
> that tourists deduce their favourable ideas of
> the general nature of the island. [11]

The London Times' San Francisco Correspondent, Donald Fraser,
wrote a series of articles on the colonies of Vancouver Island and British
Columbia with emphasis on the gold rush, finally becoming a resident in
Victoria for a period of time. Upon his arrival in June, 1858 he wrote:

> On the left is the long-looked-for Island of
> Vancouver, an irregular aggregation of hills,
> showing a sharp angular outline as they be-
> come visible in the early dawn, covered with
> the eternal pines, saving only occasional
> sunny patches of open greensward, very pretty
> and picturesque, but the hills not lofty enough
> to be very striking. The entire island, properly
> speaking, is a forest. [12]

When Fraser entered Esquimalt Harbour he was struck by its resemblance
to a Highland lake: "The whole scenery is of the Highland character.

The rocky shores, the pine trees running down to the edge of the lake, the dark foliage trembling over the glittering surface which reflected them, the surrounding hills and the death-like silence."[13] The pine trees that both he and Douglas referred to were Douglas firs, commonly called Douglas pines at that time. Then he sailed around the promontory from Esquimalt to Victoria: "The shore is irregular, somewhat bold and rocky — two more facts which confirmed the resemblance of the scenery to that of the western coast of Scotland."[14] From his wanderings around Victoria he conveyed this impression of the area:

> The scenery of the inland country round Victoria
> is a mixture of English and Scotch. Where the
> pine (they are all "Douglas" pines) prevails you
> have the good soil broken into patches by the
> croppings of rock, producing ferns, rye grass,
> and some thistles, but very few. This is the
> Scottish side of the picture. Then you come to
> the oak region; and here you have clumps, open
> glades, rows, single trees of umbrageous form,
> presenting an exact copy of English park scenery.[15]

Alexander Rattray, a naval surgeon who was based at Esquimalt for two years, authored a book on Vancouver Island and British Columbia that was published in 1862 (Figure 3,1 and 4,1). His view accorded well with that of others concerning the advantages of the Victoria area over other parts of Vancouver Island:

> The prospect from Victoria, which may be taken as
> a fair example of the general features of the island
> throughout, though fine, is by no means inviting,
> in a utilitarian point of view. It presents only a
> series of lofty, undulating, pine-clad hills, which
> rise irregularly one behind another, and have small
> intervening valleys where patches of arable land
> exist; but the hills themselves are generally so
> scantily covered with soil, as barely to afford root
> to the scattered and stunted trees on their sides and
> summits....

FIGURE 3,1
View toward the mouth of
Esquimalt Harbour circa 1860.

B.C. Archive Photo

FIGURE 4,1   Victoria circa 1860 from the south side of James Bay.

Towards Esquimalt, which lies near the hilly
country, the land is not so favourably adapted
for agricultural and pastoral farming as over the
Victoria peninsula generally. Numerous rounded
masses of rock crop out in almost every field, and
most of this neighbourhood has so broken and
rugged a character, as materially to interfere
with agricultural operations; but although much
of the land here is rocky and incapable of till-
age, it will be seen, from facts to be hereafter
stated, that what is now cultivated is highly
fertile. Towards Victoria, however, the land
improves, and in the Lake and Saanich districts
arable land is more abundant, and of good
quality.[16]

Another naval surgeon, Charles Forbes, reported his impressions
of the physical geography of Vancouver Island:

The whole country is more or less densely
wooded, excepting just where the summit of
a mountain affords no hold for plants, or where,
as in the neighbourhood of Sooke, Victoria,
Cowitchin, and Comux, limited ranges of open
grass-lands occur.... The surface is beautifully
diversified by mountain precipice, hill and
dale, widespreading lakes, and solitary tarns,
cut up by numerous arms and inlets of the sea;
in no case does the water-shed suffice to give
a navigable stream. There are no rivers, in
the stricter sense of the word, such streams as
flow through the country being simply the
short water-courses, which discharge the over-
flow of lakes or the surface-waters of the
neighbouring ridges —torrents in winter,
nearly dry in summer, valuable only as a
power for driving grist and saw mills, and possibly at a future day to be rendered useful as
a means of irrigation — a process by which
many parts of the country would be much
benefited.[17]

Forbes, along with other observers, was accustomed to the sight of navigable
rivers in England where the ria coast presented many extensive estuaries into

12

which flowed broad, slow-moving rivers. Hence, his astonishment at finding no navigable rivers on southern Vancouver Island was understandable.

One of the most authoritative writers of the time was Joseph Despard Pemberton, who was, in turn, a surveyor and engineer with the Hudson's Bay Company, a colonial surveyor, and finally, in 1860, Surveyor-General of the Colony of Vancouver Island. In his book published in 1860 he decried the plethora of inaccuracies that had been bandied about in previous writings concerning Vancouver Island and British Columbia.[18] Specifically, he singled out the 1857 article by Colonel Grant and some of the letters by the Times Correspondent, Donald Fraser, as conveying erroneous impressions of the island. However, his criticism mainly concerned the tendency of those writers to generalize for the whole of Vancouver Island on the basis of observations taken in the Victoria area.[19] But his implication is that their remarks in reference to the Victoria area were correct. Pemberton's view of the southern tip of Vancouver Island accords generally with that of others, if slightly more favourable:

> At first sight the whole country appears to be
> clothed with forest, for it is not until we travel
> inland that we ascertain that in the lowlands
> the pines take frequently the form of belts, en-
> closing rich valleys and open prairies, lawns
> in which oaks and maples, not pines, predomin-
> ate; marshes covered with long coarse grass,
> and lakes fringed with flowering shrubs, willows,
> and poplars.[20]

CLIMATE

Most visitors and settlers considered the climate to be favourable and very much like that of England. The Hon. Charles Fitzwilliam, M.P., stated: "I was in Vancouver Island in the winter of 1852-3. The climate appeared to me particularly adapted for settlement by Englishmen; it resembles the climate of England, but not quite so cold..."[21] Donald Fraser reported:

> The climate is usually represented as resembling that of England. In some respects the parallel may hold good; but there is no question that Vancouver has more steady fine weather, is far less changeable, and is on the whole milder. Two marked differences I remarked, — the heat was never sweltering, as is sometimes the case in England, and the wind never stings, as it too often does in the mother country. The climate is unquestionably superior in Vancouver.[22]

Colonel Grant, of longer residence, said, "Generally speaking, the climate is both agreeable and healthy; and not a single death that I am aware of has occurred among adults from disease during the six years that I have been acquainted with the island."[23]

Dr. Rattray conducted a detailed study of the climate of Vancouver Island, devoting a whole chapter of more than thirty pages to the subject in his book.[24] He compared the climate of Esquimalt with that of London:

> The weather of Vancouver Island is milder and steadier than that of England; the summer longer, drier, and finer, and the winter shorter and less rigorous. The mean annual temperature of the former is higher by 1.38° than that of the latter. During the summer months the hot weather of Vancouver Island is not so oppressive, and the maximum temperature is less by 14° than that of London; while in the winter the temperature never falls so low at Vancouver Island as in London; and the annual range of the two places differs by 15 1/2° in favour of Esquimalt....
> The atmosphere of Vancouver Island, especially in summer, is drier than the proverbially humid climate of England; and the heavy harvest rains

14

of the English summer and autumn months,
(especially August and September,) which
not unfrequently damage the crops, are
unknown amongst us....The table, however,
goes far enough to prove that our climate
is drier and less variable than that of England,
and that rain does not occur so often or so
heavily with us, especially in March and
April, when agricultural operations are com-
mencing, or during August and September,
when the crops are ripening....The rains of
the Vancouver Island summer have more the
character of showers. Autumn and winter
may be called emphatically the "rainy"
season, in contradistinction to the summer,
which is often excessively dry, while in
England the rain is more distributed over
the year, and summer and autumn form the
rainy season.

The prevailing winds of Vancouver Island
are southerly, those of England south-westerly;
in the former biting N.E. and E. winds, which
often prevail during the English spring and early
summer, give place to mild S. and S.W. winds
in April, which are followed in May and June
by still milder summer breezes.[25]

The table of monthly mean temperatures at Esquimalt and London presented

by Rattray probably was a fair comparison, despite the fact that the

Esquimalt figures probably were based on little more than one or two years

of record (Table 1,1). The comparison with the 30-year Standard Period

figures for Victoria indicates that Rattray's figures showed fall to be sig-

nificantly warmer and the rest of the year slightly warmer than the long

term average indicates. Hence, the temperature comparison tends to

favour Esquimalt over London to a greater extent than modern records

justify.

TABLE 1,1

MONTHLY MEAN TEMPERATURE AT ESQUIMALT, VANCOUVER ISLAND AND LONDON, ENGLAND IN DEGREES FAHRENHEIT, CIRCA 1860

Month	Esquimalt	London
January	39 (39.4)[*]	37
February	43 (41.4)	40
March	45 (44.3)	42
April	51 (49.3)	48
May	55 (54.1)	55
June	59 (57.3)	60
July	61 (60.1)	63
August	62 (60.1)	63
September	57 (57.6)	58
October	54 (51.8)	51
November	49 (44.9)	43
December	42 (41.9)	39
Year	52 (50.2)	50

*Bracketed figures are 30-Year Standard Period, 1930-1960 means for Victoria.

Source: Rattray, A. Vancouver Island and British Columbia. London: Smith and Elder, 1862, p. 49. Although it is not stated specifically, information elsewhere in the text implies that the Esquimalt figures refer to 1860-61.

Rattray exhibited considerable perspicacity in his attempt to explain the climate:

> The climate of the southern end of Vancouver Island presents certain features for which latitude alone will not entirely account.
> Its remarkable mildness, equability, and essentially insular character, are particularly noticeable; and are the combined result, first, of the equalising influence of the adjacent ocean; second, the cooling effect of the waters

16

of the Fraser River, which tends chiefly to
prevent a too elevated summer temperature;
third, the prevalence and equalising influence
of southerly winds, cool in summer and relatively warm in winter; and fourth, the rarity
of cold, damp northerly winds in summer, and
then frequent alternation during winter with
southerly currents by which their rigour is modified.

Its dryness, especially during summer, is also
remarkable, and the small annual fall of rain;
both principally the effect of contiguity to the
Olympian range, which precipitates the super-
fluous moisture of the prevailing southerly winds
before they reach the Island.[26]

Rattray correctly identified the rain shadow effect of the Olympic Mountains
and the moderating influence of the ocean, but he had an exaggerated impression of the influence of the Fraser River waters on the climate of southern
Vancouver Island.

In his discussion of climate Dr. Forbes largely concurred with the
findings of Dr. Rattray.[27] He also considered the climate as approximating
closely that of Great Britain, but he called attention to some of the except-
ional weather conditions that may occur:

The experience of the last twenty years has
shown that, at irregular periods of from five to
seven years, winters of great severity may be
expected. In 1846, 1852-3, 1859-60, and in
1861-62, the frost was intense, and the fall of
snow heavy. In neither of the two former cases,
however, was it in any way so severe as in the
latter year.[28]

Other writers, including Lieutenant Richard Mayne of H.M.S. "Plumper"
and Matthew Macfie, a Congregational minister, also discussed climate
at some length, but added relatively little new information to that provided by earlier writers.[29] Most observers tended to emphasize the sim-
ilarity with the climate of southern England, frequently commenting on

certain aspects that were considered slightly superior, and recommended
it as an admirable climate for European settlers. Although summer drought
conditions were recognized, these were not greatly emphasized as a dis-
advantage to agriculture. The climate was considered to be an especially
healthy one, at a period in history when many diseases were thought to
be more climate-related than is known to be the case today. Even in
1865 it was suggested that Victoria was a suitable place for retirement because of its favourable climate: "I know no locality so admirably suited
for ex-Indian officers and merchants to retire to— a class to which climate,
in their advanced age, is a primary consideration."[30] Undoubtedly,
reports such as Macfie's influenced many people and established the notion
at a very early date that Victoria was a place to retire to.

GEOLOGY

Detailed study of the geology of the Victoria area had not been
undertaken by the 1860's, but several observers who must have had some
knowledge of geology presented generalized reports. Two of the most
authoritative were Colonel Grant and Dr. Forbes. Grant's observations
were taken in the early 1850's when he lived at Sooke. In his paper
read before the Royal Geographical Society in 1857 he stated: "The
prevailing formation is that generally known as the gneiss and mica-schist
system: these rocks produce a broken and rugged surface, without being
attended with any picturesque effect....From Sanetch to Esquimalt gneiss
prevails, diversified with beds of dark-coloured limestone. Westwards
of Esquimalt mica slate occurs ... "[31] The only detailed study undertaken
by Grant was along the Sooke River which he reported on at some length.
Regarding the occurrence of gold, Grant considered it unlikely that gold
would be found on Vancouver Island, or at least in the Victoria area, because he said that there was very little quartz in the hills of Vancouver
Island, with which he expected gold to be associated.[32]

The report of Dr. Forbes in 1864 was probably the most accurate description of the geology at that time, and it was widely quoted by other authors:

> The geological structure of the island may be briefly described as consisting of an axis of meta-morphic gneissose rock, situated at the south-western extremity, having, resting thereon, clay-slates and other rocks, of probably Palaeozoic age.... Associated with these are lenticular masses of a semicrystalline limestone, of great economic value. On these metamorphic and trappean masses rest the sedimentary stratified rocks, which fringe the south-eastern and eastern coastlines, and which contain carboniferous strata of very great value....
>
> The most remarkable feature of the geology of the southeastern end of the island is the scooping, grooving, and scratching of the rocks by ice action....The sharp peaks of the erupted intrud-ed rocks have been broken off, and the surface smoothed and polished as well as grooved and furrowed by glacial agencies, giving the ap-pearance of rounded bosses to the numerous promontories and outlying islands which here mark and stud the coast line.
>
> As might be looked for in a country so acted on by ice, the whole surface of the land is found in this locality to be covered by boulder drift and erratic blocks of various crystalline and other rocks sufficiently hard to bear attrition. Granites and granitoid rocks of every description are to be met with; trappean rocks, micaceous schists with garnets, breccias, and conglomerates.
>
> From these granitic boulders and from the sand-stones of the outlying islands valuable building material is obtained, some of the grey granites equalling in beauty and in closeness of crystal-line texture the best granites of Aberdeen or Dartmoor.[33]

In his general statement Forbes called attention to a number of the major features of the area, including the volcanics, sedimentaries, and the abundance of glacial features, but his determination of age of certain rocks was inaccurate. The first geological reconnaissances of the Victoria area conducted by the Geological Survey of Canada were made in the 1870's by G.M. Dawson and James Richardson, the first detailed geological survey not being conducted until 1910 when C.H. Clapp carried out his survey.[34]

Much of the early interest in rock structure was centred on economic geology. Grant doubted that gold would be found on Vancouver Island, but two or three years after his comments were published placer gold was found in the Leech River valley near Victoria. Forbes drew attention to the coal-bearing sedimentaries along the east coast of the Island, extending from Comox southward to the Saanich peninsula. Of course, coal mining had been carried on at Nanaimo since the 1840's. Forbes went on to describe the coal and copper occurrences at Sooke:

> The formation, geologically, is an axis of
> trappean rock, having, resting on its north-
> western flank, clay-slates and micaceous
> schists; on its southern and south-eastern,
> a sedimentary deposit of stratified sandstones,
> shales, and seams of coal....
>   The carboniferous deposit has been proved
> by "bore" to the depth of 84 feet, and two
> thin seams of coal have been passed through.
> A promising vein of copper has been found,
> and is now being worked.[35]

Concerning Victoria and Lake districts, Forbes stated: "There are indications of copper, and a coal-seam of inferior quality crops out on the eastern coast."[36] The limestone that occurs in the Victoria area was recognized by Fraser as an important resource:"... limestone in abundance, which burns into good lime for building and for agricultural purposes..."[37] And he continued:

> A friend of mine brought down some sand from
> the sea-beach near Victoria and assayed it the
> other day. It produced gold in minute quantity,
> and I have heard of gold washings on the island.
> The copper is undeveloped. The potter's clay
> has been tested in England, and found to be
> very good.[38]

TOPOGRAPHY

Topography was illustrated on early maps of the Victoria area by
the use of hachures and shading. Edges of swamps and gullies were frequently outlined with a solid or broken line and vegetation symbols completed the landscape depiction. Following a trigonometrical survey conducted by Joseph Pemberton when he was an engineer and surveyor to the
Hudson's Bay Company, an attractive map of the southeastern districts of
Vancouver Island was published in 1855 (Figure 5,1). He made effective
use of hachures, identifying most of the prominent hills in the area between
Sooke Harbour and the northern end of the Saanich peninsula and indicating steepness of slope. The hachures were drawn with the imaginary light
source over the northwest corner of the map. Although the coastline is
reasonably accurate, the outlines of many of the lakes are very rough, for
example, Prospect, Florence and Langford lakes are misshapen, and Glen
Lake appears only as a tiny swamp.

A larger scale map of Victoria District published in 1861 indicates
a number of streams that have since disappeared as a result of urban development altering former drainage patterns (Figure 6,1). Bowker Creek,
named Tod's Stream on Pemberton's 1855 map, had two branches, one
extending northward roughly along the present day Shelbourne Street and
the other along a route south of, but paralleling, Hillside Avenue. Most
of Bowker Creek now is canalized or routed below the surface in culverts

FIGURE 5,1
A portion of the 1855 Pemberton map.

FIGURE 6,1
A portion of the 1861 Victoria District map.

B.C. Archive Photo

and the southerly branch does not appear on modern maps. Several small streams entered Victoria Harbour, most notably in James Bay, Rock Bay, and at the foot of Johnson Street, that stream running between the present Johnson and Pandora streets.

Other name changes of prominent geographic features are of interest, for example, Lake Hill became Christmas Hill, but the name has persisted as the designation of that neighbourhood, Shoal Bay became McNeill Bay and Foul Bay became Gonzales Bay, although a major thoroughfare retains the Foul Bay name. The spelling, and hence the implication of its origin, is at variance, being "Fowl" on the 1855 map and "Foul" on the 1861 map.

Topographic changes have been most pronounced in the built-up area of metropolitan Victoria. The construction of streets and highways and the installation of sewer systems has completely altered the original drainage patterns, eliminating many small streams and numerous swamps. Minor changes of slope, of course, are widespread, especially where portions of small hills have been removed for construction of roadways and buildings, sometimes by rock blasting, or where surficial deposits have been levelled. Major gravel pits at Colwood, Cordova Bay, and in Central Saanich, including the long abandoned pit on the slopes of Mount Tolmie, are prominent features of topographic alteration.

The shoreline of Victoria Harbour presents even more dramatic evidence of landscape change. Reclamation by fill has occurred particularly along the Inner Harbour shoreline where wharves and buildings have been erected on reclaimed sites. A comparison of the original shoreline as shown on the Victoria Harbour Admiralty chart published in 1861 with the present day shoreline indicates that a large part of Rock Bay has been filled in and that most of James Bay has been reclaimed, the Empress Hotel having been built on piles over what was formerly open water (Figure 7,1).

24

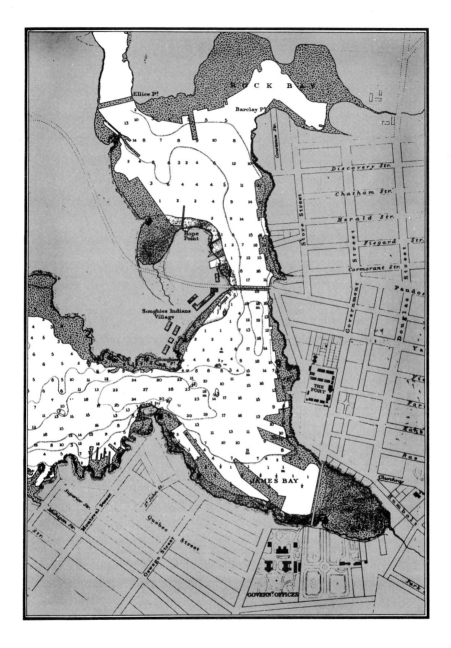

FIGURE 7,1   The present Victoria Harbour shoreline superimposed
on the 1861 Admiralty chart.

SOILS

The great variation of parent material, topography, drainage and vegetation types within the Victoria area naturally gives rise to a great variety of soils. In fact, a total of seven great soil groups are represented in the area. The first independent settler, Grant, recognized the variability of soils: "The soil under cultivation is sometimes a rich vegetable mould, in other places a clayey loam, and in others somewhat sandy."[39] In further elaboration, Forbes presented a threefold classification:

> The soils of Vancouver Island, derived from the clays and gravels of the drift that overspreads the country, and from the decomposition of the subjacent rocks, may be thus distinguished and described: —
> 1st. A poor gravelly soil, with a thin coating of vegetable mould, bearing large timber of a superior quality, coarse grass, and little underwood.
> 2nd. A calcareous loam of good quality, producing excellent crops of vegetables; very suitable for clover and other lime plants.
> 3rd. A rich dark brownish-black soil, or humus, resulting from the decay of vegetable matter, mixed in some localities with alluvium, of variable depth, and resting on a clay subsoil, over lying trap rock and concretionary limestone.
> The poverty of the soil first described, is due to its inability to retain moisture; the second is always ready for cultivation; and the third only wants subsoil drainage to carry the heaviest possible crops of wheat and of other cereals.[40]

Although he was describing the soils of Vancouver Island, all three classes occurred in the Victoria area. His first category probably refers to the brown podzolic soils of the Colwood-Langford area and the area immediately north of Elk Lake. The second perhaps relates to the well drained Acid Dark Brown Forest soils that are quite extensive on the Saanich pen-

insula and on the western sides of Saanich municipality and Victoria.
Known as the Saanichton series, these are generally the best agricultural
soils in the Victoria area. Forbes' third category likely describes the less
well drained Dark Grey Gleisolic soils represented by the Tolmie series
that is widespread on the east side of the Victoria and Saanich peninsulas.

Comments on specific areas were made by Grant and Fraser. In
speaking of the soil at Sooke, Grant probably was referring to the high
quality Saanichton Acid Dark Brown Forest soil on the north side of the
harbour: "The soil on the prairie is a rich black vegetable mould from
3 to 4 feet deep, with a stiff clay subsoil, resting on sandstone, and the
surrounding woodland also consists of very rich soil."[41] Grant mentioned
also that when cultivated the soil produced abundantly any crops that
could be grown in Scotland or England.[42] In discussing a similar type
of soil in the immediate vicinity of the town of Victoria, Fraser stated:

> The character of the soil is favourable to
> agriculture. It is composed of a black veg-
> etable mould of a foot to two feet in depth
> overlaying a hard yellow clay. The surface
> earth is very fine, pulverized, and sandy,
> quite black and, no doubt, of good quality;
> when sharpened with sheep-feeding it pro-
> duces heavy crops... In some places the
> spontaneous vegetation testifies to the rich-
> ness of the soil,— such as wild pease or
> vetches, and wild clover, which I have seen
> reach up to my horse's belly,— and a most
> luxuriant growth of underwood, brambles,
> fern, etc.[43]

The many swampy areas in Victoria District also were regarded as
potentially favourable for agriculture by Forbes.[44] Concerning the need
for drainage improvement, Rattray quoted one of the farmers of the colony:
"Draining is unknown in Vancouver Island as yet; labour is so high that it
prevents the farmer from carrying on this improvement. Could he afford

the luxury, the soil would be thereby rendered equal, if not superior to

that of England and Scotland. Our best soil is low and swampy, requiring

nothing but cheap labour and tile-drains to make it equally productive

with the finest lands in the old country."[45]

Although no official soil surveys were carried out, most of the

early settlers recognized the dark brown and black soils of the area as the

highest quality for crop production. Also, however, they tended to favour

the more open areas, or "prairies" as they called them, for cultivation,

in order to avoid the great effort of clearing dense forests. Probably in

reference to the Saanich peninsula, Macfie commented, "In a district

about a dozen miles from Victoria I have seen a single prairie containing

not less than 400 acres of clear land where the alluvial soil, consisting

mainly of black loam, was at least a couple of feet thick."[46] Hence,

it was the better soils supporting grass and scattered trees that were cul-

tivated initially, followed in later years by the clearing for agriculture

of more densely wooded lands. In retrospect, it is unfortunate that urban

development subsequently spread over a large part of the most productive

soils in the Victoria area.

VEGETATION

In reference to the state of the landscape in the mid 1850's,

Colonel Grant asserted:

> The lands, at present surveyed by the Hudson
> Bay Company are included in a line, which may
> be taken from Sanetch to Soke Harbours; the
> quantity of land surveyed in detail is 200 square
> miles, of which one-third is rock or unavailable,
> the remainder is principally woodland. The pro-
> portion of open land will be seen from the above
> remarks, where all that is known is mentioned,
> and bears a very small proportion to the wood-
> land; but where it exists at all it is almost in-
> variably rich; and the woodland, where it is

28

at all level, is richer than the prairie ground,
from the increased quantity of vegetable deposit.[47]

The "prairies" referred to were more parklands, with scattered Garry oak
trees interspersed with grasses, clovers, and camass, as well as other small
plants (Figure 8,1). It is known today that Garry oak was the original
vegetation on dry sites on the Victoria and Saanich peninsulas, and Arbutus and Douglas fir on moist sites.[48] The fact that the "prairies" were
more extensive than would have been the case if they had occurred only
on dry sites may have been due to the removal of other species, principally Douglas fir, on some of the deeper, moist soils in the pre-colonial
period. Many observers, including Grant, commented on the fact that
the Indians had a habit of setting fire to the woods in summer, perhaps to
clear areas for their potato planting, as well as to increase the extensive-
ness of the camass in grassy areas. Alluding to the camass, Grant stated:

> The open prairie-ground, as well as the patches
> of soil which are met with in the clefts of the
> hills, are principally covered with the camass,
> a small esculent root about the size of an onion,
> with a light-blue flower, the Camassia esculenta
> of botanists. The camass constitutes a favourite
> article of food with the savages, and they lay up
> large quantities of it for winter consumption, bury-
> ing it in pits in the ground in the same way as they
> keep potatoes.[49]

Considering the various districts and the character of their vegetation,
Forbes stated the following concerning Sooke: "All over the surrounding
broken country there is excellent grazing, during seven months of the year,
the wild vetch growing luxuriantly to a height of three or four feet. On
Sooke River there are many fine, though limited valleys, all bearing magnificent timber, cedar especially."[50] And of Metchosin he said: "There
is some fine grazing-land, but little prairie, heavy timber covering the whole.

FIGURE 8,1
Oak parkland around Fort Victoria
in the late 1850's.

B.C. Archive Photo

The pines are very fine, but far back from the sea. The whole district is very beautiful and salubrious, well sheltered, with a dry gravelly soil, adorned with Druid-like groves of oak and solemn-looking clumps of pine, intermingled with the varied foliage of a thick shrubby undergrowth."[51] Of Esquimalt Land District he stated: "The soil, generally, is poor in quality, covered with scrubby timber, a great deal of rock, and many lakes and large swamps."[52] Victoria District he considered more favour-able: "The whole surface is undulating: in most places thickly timbered, in the neighbourhood of Victoria clear, and sweeping along the coast-line as fertile grassy pastures."[53] The Saanich and Lake districts were described as similar in character to Victoria District.[54]

Of course, it was the Douglas fir that was of greatest economic interest. Macfie asserted: "It is now universally admitted that Vancouver Island and British Columbia produce the best qualities of timber to be found in the world....This tree is in great demand for spars; and for strength, lightness, elasticity, erectness, beauty of grain, and height, it cannot be surpassed."[55] He went on to say that a Douglas fir spar more than 200 feet high had been erected in Kew Gardens, London and that sections from a tree 309 feet long had been sent to England for the International Exhibition of 1862.[56] Both Macfie and Rattray presented extensive lists of the major trees of the Victoria area. Rattray stated:

> The Douglas pine preponderates at the southern
> end of Vancouver Island, and along its east and
> west coasts, with occasional patches of oak, and
> a few maple, cypress, arbutus, yew, and other
> varieties....Many of the trees on the hilly ground
> are of stunted growth; but in the valleys and low
> ground, especially along the west coast, heavy
> timber is plentiful — especially the lofty Douglas
> pine, admirably adapted for mast and spar
> making....Much of the oak of this colony is of
> good size and quality, and well adapted for
> knee-timber and general ship-building purposes.[57]

The grasses occurring with scattered trees, which led to the designation of "prairies", were associated with many other small plants considered of economic value. Fraser reported that the farmers of the Victoria area depended on the natural grass as food for their cattle and sheep.[58] Pemberton commented in more detail on the small plants:

> The native grasses of the country are of a poor
> Alpine character, springing up early in April
> and dying away in September; swamp grass ex-
> cepted, which supports the stock of the country
> in winter, but which is too coarse and woody
> in the fibre to fatten them. This deficiency is,
> however, to a great extent counterbalanced by
> native tares, clover, and vetches, which are, in

31

most localities, abundant. The open grounds,
also, grow berries of many kinds, and roots,
such as onions, kamass, etc., on which the
Indian, to a great extent, subsists.

In many places the wild flowers of England,
and the common garden flowers, such as lilies,
lupins, orchids, etc., occur in profusion, and
blossoming shrubs of infinite variety.[59]

A number of large scale maps produced between 1855 and 1861
yield a magnificent insight into the distribution of basic vegetation types.
Symbols on the land district maps of 1858 and 1859 clearly distinguish between coniferous and deciduous trees, and the placement of the symbols
appears to indicate open areas where the land was either under cultivation
or mainly non-forested. Poorly drained or swampy areas also are indicated.
The symbols are roughly applied, as the maps were not precisely drafted
for publication, and explanatory legends were not included. The map of
the Victoria District, however, was redrafted in a more sophisticated
version published in 1861, but still without a legend (Figure 6,1). The
1855 Pemberton map is useful as an aid to the interpretation of the land
district maps (Figure 5,1). It indicates swampy and poorly drained land,
as well as the major breaks between open and wooded land. The maps
permit the separation of the Garry oak-grassland community from those
dominated by coniferous species and the indentification within the deciduous group of poorly drained areas containing swamp land with scrub
vegetation on the margins.

An 1860 vegetation map of Victoria District has been prepared,
based mainly on the Victoria Land District maps of 1858 and 1861, with
wooded area boundaries being checked by use of the 1855 Pemberton map
(Figure 9,1). That land district today is almost completely urbanized and
most of the original vegetation cover has been obliterated. Even by 1860
an extensive area of streets and lots had been plotted, but probably was
not completely cleared, especially the five acre lots on the north side of

32

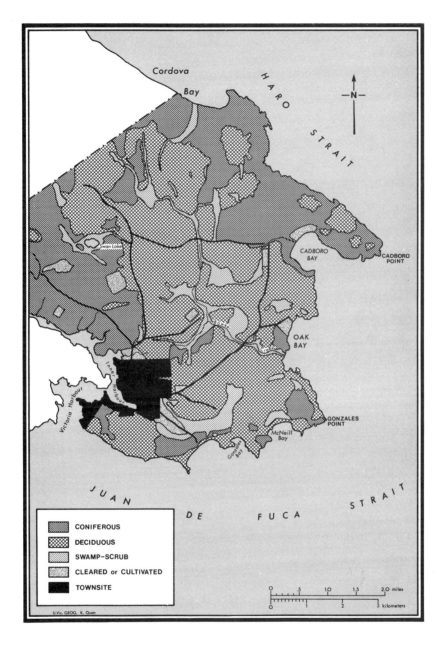

FIGURE 9,1   Victoria District vegetation map, 1860.

33

the town. However, the maps do not indicate vegetation types within the townsite. One can infer from the surrounding vegetation and from early reports, especially those of James Douglas, that most of the townsite was of the oak-grassland community, with coniferous trees along the northern shore of the James Bay district in the vicinity of Laurel and Shoal points.

It is apparent that the oak-grassland community dominated the Victoria District, with the exception of a broad belt of coniferous in northeastern Saanich from Cadboro Point to Cordova Bay and another zone in southwestern Saanich along the Gorge. Outside those areas a few small patches of coniferous existed in Victoria city and Oak Bay municipality. The scrub vegetation in the poorly drained areas was quite widespread in distribution, particularly along Bowker Creek and its tributaries, in the Fairfield district, and around Swan Lake. Although the extensive "prairie" areas have succumbed to urban development, some of the parks and golf courses give an impression of the parkland type of vegetation, particularly Uplands Park in Oak Bay. The coniferous vegetation type may be observed in Mount Douglas Park and on the University of Victoria campus. Most of the cleared areas were adjacent to the town or represented the sites of well documented early farms, including Craigflower, Hillside and Uplands. The sizeable cleared areas at Cadboro Bay and Oak Bay probably contained farms, but also are known to have been sites of winter villages of the Indians in pre-colonial times.

A Saanich peninsula vegetation map of 1860 has been produced from the land district maps of the peninsula (Figure 10,1). The North and South Saanich sheets seem to be preliminary drafts similar to the 1858 Victoria District map, with the same symbolization. The Lake District map, on the other hand, represents a higher standard of cartographic a-chievement that was lithographed at the Topographical Depot of the War Office in London (Figure 11,1). It was one of a series of land district maps, the Shawnigan and Quamichan sheets being other examples, pro-

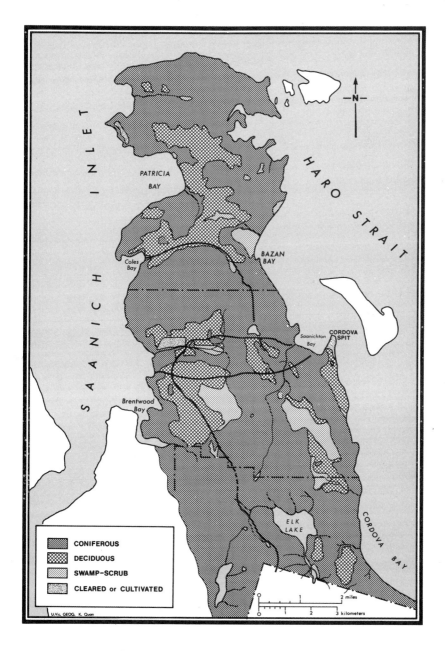

FIGURE 10,1   Saanich Peninsula vegetation map, 1860.

CONIFEROUS
DECIDUOUS
SWAMP-SCRUB
CLEARED or CULTIVATED

U.Vic. GEOG. K. Quan

35

FIGURE 11,1
A portion of the 1859 Lake District map.

duced as specific guides to prospective settlers. Detailed hachuring was used to indicate topography and notations were spread across various features of the map to give localized information on quality of land and vegetation. These site notations included such comments as, "good land", "light soil", "land of medium quality" and "good soil much broken by rocky knolls and deep ravines". Agricultural and urban development today occupies most of the lowland areas of the Saanich peninsula, but upland areas still retain evidence of the original vegetation.

Coniferous trees dominated the peninsula, especially at the higher elevations. Deciduous vegetation, including grassland, was most extensive in the Brentwood area of South Saanich Land District and in the Coles Bay, Victoria International Airport and Patricia Bay areas of North Saanich. Poorly drained areas of swamp and scrub were widespread, but of greatest extent east of Brentwood in two large areas that, in recent years, have been devoted substantially to crops of potatoes. Few cleared areas were indicated, the only one definitely resembling a farm being at Hagan Creek north of Brentwood. The small clearings at Coles Bay and Patricia Bay were Indian village sites. The other clearings may have been due to fire or other natural causes. Certainly, Cordova Spit at Saanichton Bay was too sandy to support much vegetation. The Lake District map was the only one that specifically identified "prairie" areas. With the exception of the small area immediately east of Elk Lake, they were associated with scattered deciduous trees, consistent with their occurrence on the other map sheets. Near Elk Lake, however, scattered coniferous symbols appear on the "prairie" area, indicating that grassland sometimes was associated mainly with conifers.

CONCLUSION

This study represents an attempt to turn back the clock more than a century in order to depict a physical landscape that in its entirety no longer exists. Vegetation, drainage, soils and topography have been considerably altered within the metropolitan area. In the case of drainage and topography it is relatively easy to document the changes with ordinary topographic maps. Although there was no soil survey, there was a valuable legacy in the form of a series of maps showing generalized vegetation types, from which certain soil characteristics can be inferred. The 1860 vegetation maps provide valuable information for the historian attempting to explain why particular sites were chosen for farms or other specific purposes.

Of even greater importance to the understanding of historical events was the manner in which the early residents and visitors perceived the physical landscape. Their interest in the area was strongly dominated by economic realities: rock structure was important to them as a guide to the occurrence of valuable minerals, a knowledge of soils, topography and vegetation was important for the development of agriculture, and forests were important for the production of spars, timber and lumber. But they also looked at the landscape with political and esthetic considerations in mind. The Hudson's Bay Company found Victoria Harbour quite advantageous as a fur trading headquarters for the Pacific coastal region and the British Navy found Esquimalt Harbour most suitable as a naval base. Many writers envisioned Victoria as a potential centre of trade and commerce in the Pacific, eventually rivalling San Francisco, and called Vancouver Island the England of the Pacific. The favourable climate was looked upon as providing an attractive locale for English settlers, even those in retirement under genteel circumstances.

Particularly noteworthy was the constant comparison with England and Scotland. The striking resemblance of the climate and agricultural possibilities to those of the Mother Land was emphasized by most observers. Many writers, in fact, drew a parallel between Victoria and

38

London, often to the advantage of Victoria. The climate was considered better in some respects and the forests were considered superior for lumber and timber. The oak parkland of Vancouver Island that so resembled the landscape of England was like an oasis in an alien land to many an English traveller. There was a recognition of many familiar plants that were similar to species in England. It was appreciated with satisfaction that many of the flowers, vegetables and fruits grown in England could be successfully raised in the Victoria area. From the early days of its settlement Victoria was earmarked as a special place among the far flung possessions of the British Empire where Englishmen and Scotsmen could feel a little more at home than elsewhere. Undoubtedly, this stamp imparted in the 1860's was of long lasting importance, continuing down to the present day as a recognizable characteristic of the region. Elements of Victoria's "bit of old England" tourist image and its reputation as a "good place for an Englishman to retire to" may be traced back to the prevalent perception of the physical geography around 1860.

REFERENCES

1. ROBINSON, L.B. Esquimalt. Victoria: Quality Press, 1948, p. 28.

2. Ibid.

3. GRANT, W.C. "Description of Vancouver Island," Journal of the Royal Geographical Society, 27 (1857), p. 290.

4. DOUGLAS, J. "Report to John McLoughlin, July 12, 1842," as quoted in PETHICK, D. James Douglas: Servant of Two Empires. Vancouver: Mitchell Press, 1969, p. 53.

5. Ibid.

6. Ibid.

7. Ibid., p. 54.

8. Ibid., pp. 54 and 55.

9. Ibid., p. 55

10. GRANT, W.C., op. cit., p. 269.

11. Ibid., pp. 271 and 272.

12. FRASER, D. "A Trip to Vancouver", London Times, August 27, 1858, as quoted in HAZLITT, British Columbia and Vancouver Island. London: G. Routledge, 1858, p. 204.

13. Ibid., p. 206.

14. Ibid., p. 209.

15. Ibid., p. 217.

16. RATTRAY, A. Vancouver Island and British Columbia. London: Smith and Elder, 1862, pp. 58 and 59.

17. FORBES, C. "Notes on the Physical Geography of Vancouver

Island", Journal of the Royal Geographical Society, 34 (1864), p. 155.

18. PEMBERTON, J.D. Facts and Figures Relating to Vancouver Island and British Columbia. London: Longman, Green, Longman and Roberts, 1860.

19. Ibid., p. 3.

20. Ibid., p. 9.

21. FITZWILLIAM, C., as quoted in HAZLITT, W.C., op.cit., p. 200.

22. FRASER, D., as quoted in HAZLITT, W.C., op.cit., pp. 217 and 218.

23. GRANT, W.C., op.cit., p. 275.

24. RATTRAY, A., op.cit., pp. 22-55.

25. Ibid., pp. 48-51.

26. Ibid., pp. 41 and 42.

27. FORBES, C., op.cit., pp. 157-160.

28. Ibid., p. 159.

29. MAYNE, R.C. Four years in British Columbia and Vancouver Island. London: W. Clowes and Sons, circa 1863, pp. 422-425. MACFIE, M. Vancouver Island and British Columbia. London: Longman, Green, Longman, Roberts and Green, 1865, pp. 174-182.

30. MACFIE, M., op.cit., p. 182.

31. GRANT, W.C., op.cit., p. 270.

32. GRANT, W.C. "Remarks on Vancouver Island, Principally Concerning Townsites and Native Population," Journal of the Royal Geographical Society, 31 (1861), p. 213.

33. FORBES, C., op.cit., pp. 155 and 156.

34. CLAPP, C.H. Geology of the Victoria and Saanich Map-Areas,
        Vancouver Island, B.C., Memoir 36, Ottawa:
        Geological Survey of Canada, 1913, p. 3.

35. FORBES, C., op.cit., pp. 160 and 161.

36. Ibid., p. 162.

37. FRASER, D., as quoted in HAZLITT, W.C., op.cit., p. 214.

38. Ibid.

39. GRANT, W.C., op.cit., (1857), p. 274.

40. FORBES, C., op.cit., p. 156.

41. GRANT, W.C., op.cit., (1857), p. 284.

42. Ibid., p. 283.

43. FRASER, D., as quoted in HAZLITT, W.C., op.cit., pp. 214 and
        215.

44. FORBES, C., op.cit., p. 161.

45. RATTRAY, A., op.cit., pp. 60 and 61.

46. MACFIE, M., op.cit., p. 183.

47. GRANT, W.C., op.cit., (1857), p. 289.

48. MCMINN, R.G., EIS, S. and OSWALD, E. "Native Vegetation,"
        An Inventory of Land Resources and Resource Potentials.
        Victoria: Capital Regional District, 1973, p. 59.

49. GRANT, W.C., op.cit., (1857), p. 289.

50. FORBES, C., op.cit., p. 161.

51. Ibid.

52. Ibid.

53. Ibid.

54. Ibid., p. 162.

55. MACFIE, M., op.cit., pp. 131 and 132.

56. Ibid., p. 132.

57. RATTRAY, A., op.cit., pp. 74 and 75.

58. FRASER, D., as quoted in HAZLITT, op.cit., p. 215.

59. PEMBERTON, J.D., op.cit., p. 19.

PLATE 3
Urban development on marine sediments.

44

CHAPTER 2
HUMANISING THE URBAN FOREST[1]

Michael C.R. Edgell

University of Victoria

In a volume devoted to the character and management of the physical environments of Greater Victoria, the title of this chapter may appear as something of an anomaly. "Humanising" and "urban" have little physical connotation, and "forest" may seem to be a biological rather than physical afterthought. But where is the boundary between the physical and biological components of environment; and do not most of us live in a suburban, if not urban environment? We pass our daily lives and fill our leisure time in environments that we ourselves have created. No one disputes the origin of our urban landscapes. Yet no less a product of human activity are those 'natural' playgrounds, landscapes and forests surrounding the city. The nature that awaits the escaping urban dweller in the diverse rural and forest landscapes an hour's drive from home is, in fact, man-made. It is a local manifestation of what on a global scale has been called the "humanising of the earth" by Rene Dubos;[2] a process in which man has changed the physical and biological characteristics of the land and thereby added "human order and fantasy to the ecological determinism of nature."[3]

Perhaps in no other component of the landscape complex are the results of this humanising so visibly striking as they are in the vegetation. Contrast, for example, the relative uniformity, even monotony, of the few remaining tracts of undisturbed forest on southern Vancouver Island with the diversity of landscapes and vegetation in more settled areas such as the Greater Victoria region (Figure 1,2). In the latter area, distribution of vegetation, transitions from forest-type to forest-type and

FIGURE 1,2
Forest and Field, Saanich Peninsula.

from forest to non-forest, together with many attributes of vegetation dynamics and composition, owe as much to man's activities as they do to biological and physical factors (Figure 2,2 and 3,2). In the sense that such 'cultural' landscapes and vegetation have been transformed from nature by man, they are artificial. But artificiality does not imply inferiority or that these areas are necessarily ecologically depauperate or unstable compared with 'natural' environments. Indeed, many of the economic, aesthetic, psychic and physical needs of man are better satisfied by such humanised environments than they are by raw nature.

But if artificial humanised landscapes are not inferior to natural ones for many purposes, the maintenance of their values and viability certainly depends upon continued active care and management. Having evolved with man, these environments can only be maintained by him. Society can also feed new values and needs into management alternatives and decisions, and in this way, environments and landscapes can be managed not only to maintain economic, ecologic and psychic values, but to evolve and respond to satisfy new social objectives.

One of the areas where such response is particularly marked, and yet still most needed, is in and around urban centres. Rapidly growing populations; burgeoning and changing demands for recreation outlets and facilities; increasing conflicts for the use of limited resources or particular areas; questioning of established tenets of 'single' and 'multiple' use, and growing concern with the quality of our daily environments; these factors and more compound the problems of landscape management. In the suburban forest fringes and forested lands within easy reach of urban dwellers, the so-called 'urban forests', it is true that "every patch of woods and sometimes every tree [can become] a bone of contention and a management problem."[4]

This chapter attempts to provide an overview of some of these problems and the potentials for forest land management within the specific

47

FIGURE 2,2
Urban forests for people to enjoy.

FIGURE 3,2
Road edges can enhance diversity.

context of the urban forests of Greater Victoria. These forests currently provide a wide range of services for the area's population. But we will enquire if these already humanised environments can be yet further managed and humanised to better and more completely satisfy the needs of that urban population.

THE URBAN FOREST:
DEFINITION AND POTENTIAL

The growing impacts and demands of urbanisation upon forest environments in North America have recently given rise to a small but distinct field of urban forestry. Major emphases in this field are the need to rethink traditional viewpoints of forest management and to examine the roles that forest environments can play in satisfying various psychological, social and physical needs of the urban dweller.[5] Some of this thinking has drawn inspiration from the history and success of intensive multiple-use forestry in highly urbanised western and central Europe. But a more direct impetus has been provided by conditions in the eastern provinces and states of North America. Massive encroachment of urban areas into forest lands, the plight of woodlot, park or even individual tree 'leftovers' in urban areas, and the readvance of forests over large areas of abandoned farmland close to cities, have provided the necessity and opportunity to think in terms of an "urban forest".

Although initially the concept of the urban forest was limited to wooded areas within the urban areas, it has now broadened to encompass also the "peri-urban" or urban hinterland areas. Simply defined therefore, urban forests are those that are utilized and influenced by urban populations. Rather than think in terms of an urban forest, it is better to visualise an "urban forest region" encompassing watersheds and recreational areas serving the urban zone, the suburban fringe and intervening rural areas, and the urban centre itself.[6] Whilst such a region is difficult to define

49

spatially with any accuracy, and will obviously vary from area to area, it is a well defined region functionally, serving the diverse needs of the urban population (Figure 4,2). It provides, for example: environmental regulation in hydrological and soil processes; environmental amelioration such as screens for dust, air pollution, wind and noise; aesthetic values in both tree dominated and urban landscapes; a setting for a wide range of passive and active recreation activities, and a sanctuary for wildlife. Trees can also significantly increase land property values.[7]

In spite of the pressures exerted on them and the need to accomodate those pressures, urban forests are perhaps not the place where simultaneous multiple use can be developed to the fullest extent. Public health may apparently prevent recreational access to municipal watersheds; public safety precludes hunting in most areas. Aesthetics or preservationist sentiment may prevent logging in some areas, but in others, timber production may help to defray the costs of silvicultural management for recreational purposes. But if we accept the need to actively manage forest areas for optimum social benefit, then in urban as against wilderness areas, a much greater range of practices is appropriate, possible and necessary.

Urban hinterlands therefore present some of the most challenging opportunities and potentials for integrating the forest environment into modern social needs; not only to preserve and protect forests, but also to bring the many values of forest landscapes into peoples' everyday lives and experiences. Evidence is overwhelming in its insistence that the greatest need for the planning of quality and satisfaction in our environments is where people live. This is not to deny the continued need for improved environmental management in wilderness or wildland areas. But few of these are situated close enough to cities to satisfy the immediate or mundane day-to-day and weekend needs of urban dwellers. And however urban growth is controlled or located, it will continue to impose

FIGURE 4,2
A view of the urban forest.

heavy demands on, and also require intensified management of, forests within and just beyond the urban fringe.

It is not enough to set aside forested open spaces in the forest-suburban fringe on a negative basis of low utility; so often such areas become ecological and aesthetic slums. It is also insufficient to reserve land or provide recreation facilities with no regard or planning for the type of recreation that urban dwellers prefer and desire. Areas should be chosen and managed for their positive recreational, amenity and ecological values relative to residents' needs and the regional resource base. We should also more closely examine the value of retaining forested landscapes as locations for planned residential development.

We tend to think in a dichotomous manner of "urban" and "rural", "city" and "forest". We bemoan the fate of landscapes, forests, resources and solitude strangled by the tentacles of urbanisation. But can we not also plan for the possibility that forest and city can be integrated, each contributing to the social viability of the other? This is the challenge and potential facing us in the management of the urban forest; to improve the humanised living environment of urban man.

VICTORIA'S URBAN FOREST

Forest Characteristics

Forest areas of varying character and size still occupy a significant proportion of the Greater Victoria area, and all of them lie within an hour's drive of any location within the metropolitan region. A broad almost continuous forested belt to the west and north-west of the main urban areas marks the southeastern edge of the more extensive forests of Vancouver Island (Figure 5,2). To the north, in the Saanich Peninsula, disjunct forest blocks are interspersed in a rural agricultural-suburban matrix; whilst within the urban region, a number of parks and residual woodlots provide a scattering of small forested areas. These varied forest

52

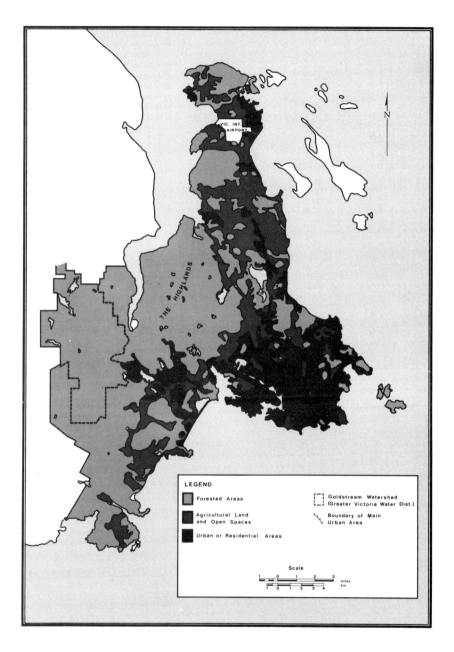

FIGURE 5,2   Generalized land-use categories of Greater Victoria.

landscapes have evolved through a complex interplay of environment, vegetation dynamics and land use. The imprint of man upon their development and character is apparent everywhere. Much of their diversity, utility, ecology and aesthetic quality stem from that imprint.

The original forests in this mainly summer-dry climatic region consisted dominantly of western red cedar (Thuja plicata), grand fir (Abies grandis), Sitka spruce (Picea sitchensis) and red alder (Alnus rubra) on wet sites, grading through Douglas-fir (Pseadotsuga menziesii), grand fir or mixed fir-arbutus (Arbutus menziesii) on moist and dry sites, to Garry oak (Quercus garryana) and arbutus on more xeric areas. The Indians found the relatively open forests attractive for settlement and probably modified some of them through burning and digging during the collection of camas bulbs (Camassia quamash) and the cultivation of potatoes. Human impact on the forests intensified following the establishment of Fort Victoria and the arrival of the European colonists.[8] In the past 130 years a history of agricultural and residential clearing, livestock grazing, early high grade logging and later selective or clearcut felling often followed by fire, has transformed the early forestscape. Seral (regrowth) communities now occupy practically all the original forest areas (Figure 6,2), the major exceptions being limited areas of the Goldstream watershed that still contain mature stands, and a few scattered fragments in, for example, Mount Douglas and John Dean Parks.

Detailed description of the floristic and ecological characteristics of these communities is extraneous to the major theme of this chapter. However, it is useful to group them into four main forest-types (Figure 7,2).[9] Whilst the general characteristics within each of these types are uniform, there are many local variants due to the interaction of successional development and site conditions.

Garry oak woodlands

Communities dominated by Garry oak were probably once more

FIGURE 6,2
Mosaics of regrowth forest dominate.

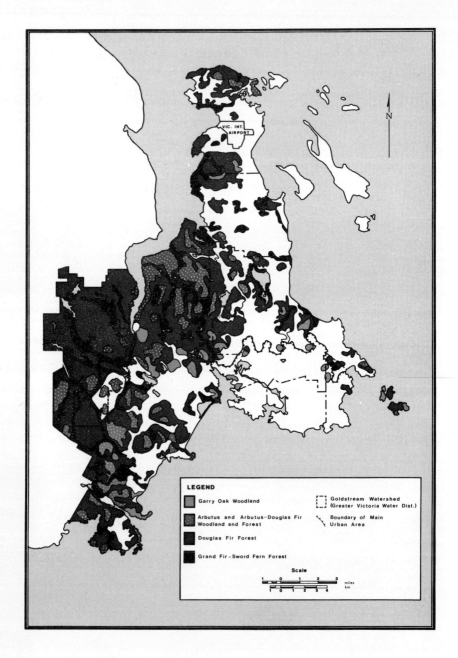

LEGEND

Garry Oak Woodland

Arbutus and Arbutus-Douglas Fir
Woodland and Forest

Douglas Fir Forest

Grand Fir-Sword Fern Forest

Goldstream Watershed
(Greater Victoria Water Dist.)

Boundary of Main
Urban Area

Scale

FIGURE 7,2   Simplified forest types of Greater Victoria.

56

widespread than now, but have been largely obliterated or extensively altered because they occupied sites that were later to become prime residential areas. Yet the European settlers may paradoxically have initially increased the relative abundance of oak in many areas through the selective removal of other (coniferous) trees. Reminding the settlers of European oaks, the local species was probably favoured in early landclearing and landscaping. The results of this early stage are represented in the few vestiges of "oak parkland" that dot the Victoria landscape (Figure 8,2). Extensive stands of oak are now restricted to residual pieces of scrub woodland on thin-soiled rock and upland areas. The often attractive twisted forms of the oaks on these poor sites, in conjunction with open rock areas, extensive views into the open woodland, and a colourful show of diverse spring flowering bulbs, provide high aesthetic and conservation values (Figure 9,2). Many areas are highly regarded as residential sites, although they are of no value for commercial timber production. For any but the lowest intensity use, these fragile sites have a limited carrying capacity, although a wider range of uses is possible if some degradation, especially in the ground flora, is acceptable.

Arbutus-Douglas fir woodland and forest

Arbutus, along with Garry oak, is the most characteristic, and due to its colouration and evergreen broadleaf form, the most striking forest tree in the region (Figure 10,2). Even more than oak, its range and density seem to have been increased by human activity through the creation of open conditions and the removal of competition. It grows in a varying matrix with Douglas-fir over a wide range of rocky and/or freely drained sites, (Figures 11,2 and 12,2), and like oak, also grows well on deeper moister soils. In some coastal locations it forms almost pure stands. The relative abundance of arbutus and fir depends largely upon the openness of the site and the successional stage of the community.

FIGURE 8,2
Oak parkland, Uplands residential area.

FIGURE 9,2
Open oakwoods survive in rocky areas.

FIGURE 10,2
The forests' appeal owes much to Arbutus.

FIGURE 11,2
Mixed Arbutus-fir, Saanich Inlet.

FIGURE 12,2
Arbutus can form dense stands.

Open poor-growth sites tend to favour arbutus as it suffers less competition from Douglas-fir. Similarly, arbutus is more frequent in early seral stages following logging, fire or clearing, when it often develops into dense thickets, either from seeds dispersed mainly by pigeons, or from cut stumps which generate coppice shoots. Conversely, as the community develops, Douglas-fir gains a competitive advantage, and the frequency of arbutus declines. The shrub and ground floras are more diverse than those of many oak communities, with which the present ones intermix. Generally, arbutus-fir woodlands are also visually more complex than oak woodlands due to the mix of trees and differences in life forms. At higher elevations, this diversity is enhanced by the appearance of lodge-pole pine (Pinus contorta) and manzanita (Arctostaphylos columbiana) (Figure 13,2). Even more than the oak communities, arbutus-fir areas are favoured residential sites (Figure 14,2). Thin soils and steep topography, however, often make servicing difficult, and also curtail the capacity for sustaining anything more than moderate recreational use.

Douglas-fir forests

These are the most widespread forest types in the area, occupying terrain ranging from moderate slopes mantled by glacial till or stoney colluvium to level sites blanketed with coarse gravels. The forest varies from a relatively open one of Douglas-fir with a salal (Gaultheria shallon) and Oregon grape (Berberis nervosa) shrub layer, to thicker forests containing mixed Douglas and grand fir with denser and often moss-rich understories on deeper soils or north-facing slopes (Figure 15,2). Residual arbutus is frequent in some stands that appear to be later successional stages of arbutus-fir communities. Compared with the preceeding oak-arbutus-fir communities, Douglas-fir forests have a greater and more uniform tree density. There is a certain visual monotony in the spacing of straight trunks and the dense shrub layers, and a lack of floristic diver-

FIGURE 13,2
Arbutus and pine fringe upland viewpoint.

FIGURE 14,2
Arbutus and fir frame new housing.

FIGURE 16,2
Veteran firs stand above regrowth.

FIGURE 15,2
Open Douglas fir forest.

sity. But the larger trees have an aesthetic appeal of their own (Figure 16,2). Some stands have moderately good growth rates, and have the potential for limited commercial timber production. When opened up to residential or other clearing, however, the previously closed stands become prone to windthrow. Although less attractive, diverse and traversable than the more open woodlands, Douglas-fir forests are generally more tolerant of moderate to heavy use for a variety of purposes.

Grand fir-cedar-swordfern forests

Occupying low-lying and highly productive sites characterised by deep soils and permanent or seasonal water seepage, these forests have been significantly altered by logging and agriculture. But many areas retain trees of impressive stature which, in conjunction with often lush swordfern (Polystichum munitum) and moss layers, provide a unique aesthetic appeal in contrast to the more open and lower forests of drier areas (Figure 17,2). The largest trees are usually veteran Douglas-fir, but grand fir and western red cedar dominate many stands, and western hemlock (Tsuga heterophylla) may be frequent. Bigleaf maple (Acer macrophyllum) and Pacific dogwood (Cornus nuttallii), are common, and red alder characteristically dominates stands in early successional stages following disturbance (Figure 18,2). In topographic depressions and lake basins with stagnant drainage and organic soils, these forests merge into stands of cottonwood (Populus trichocarpa), western red cedar, western hemlock, alder and willow (Salix spp.). These latter stands are small in extent, and along with associated wetland communities, such as Sphagnum bogs with lodgepole pine, have not been included in Figure 7,2 .

Considered together, these four major forest-types constitute an ecologically, economically and aesthetically varied assemblage, with a marked regional character differentiating them from forests further west and north on Vancouver Island. A summary of their more notable attributes

FIGURE 18,2
Alder regrowth on disturbed wet site.

FIGURE 17,2
Mixed wet site forest.

would include:

1.　　Large differences in age classes, tree sizes and stand densities due largely to past logging history;

2.　　A variety of life forms including needleleaf evergreen, scaleleaf evergreen, broadleaf evergreen and broadleaf deciduous;

3.　　A wide mix of tree species;

4.　　Great variations in individual tree growth forms, from gnarled twisted specimens to tall, straight-stemmed individuals, dependant on site diversity and past logging, burning and clearing;

5.　　An abundance of undershrubs and colourful spring flowering bulbs which are rare or absent in surrounding areas;

6.　　Many varied edge and contrast zones between forests and rural, pastoral, residential, marine coast, lakeshore, stream bank, wetland and rock outcrop areas;

7.　　An abundance of viewpoints that provide wide scenic vistas of forest/upland/water constrasts and also allow extensive views into and through the immediate forest foreground;

8.　　A dynamic ecology, based on the fact that most communities are seral, that can significantly alter the attributes noted above.

Judged by the present quality, diversity, aesthetics, wildlife habitat and recreational opportunities of these forests, the impact of past 'destructive' high grade selective and clear cut logging was not entirely detrimental (Figure 19,2 and 20,2). Research elsewhere has suggested that to many people, an aesthetically superior forest landscape is characterised by variety in both background scenery and foreground composition. Such a landscape contains a species mix of uneven-aged (uneven-sized) well-formed healthy trees, with occasional small openings and pure even-aged stands and some malformed (picturesque) trees, including dead ones in various stages of decay.[10] These attributes, due largely to past logging and related activities, characterise many of the

66

forest types and landscapes that dominate the Victorian urban forest. The juxtaposition of agricultural and residential areas with forests also adds diversity, creating obviously humanised landscapes within which the city dweller can feel comfortable, to balance the apparently less humanised, more natural environments into which he can venture to "create an atmosphere of harmony between him and the rest of creation."[11] Such a varied forest cover provides a superb setting and potential for a wide variety of activities, experiences and facilities for the urban population. But much of this potential is currently untapped, and it is therefore in the planning and use, rather than in the character of these forests, that any inadequancies may exist.

Forest Uses

Private land accounts for the bulk of the urban forest, and the amount of public land over which direct land use controls can be exerted in provincial, regional and municipal parks is relatively small (Figure 25,2). However, the 15,000 acre Goldstream watershed, part of the 33,000 acre Greater Victoria Water District catchment, must also be considered potential public land in the urban forest, although currently closed to public access. In the southern Highlands, 3,500 acres recently designated for future urban development and partly purchased by the provincial government, could also be considered a special category of public land.

A limited amount of contract logging is carried out on private lands in the Highlands district, but logging in general is no longer an important regional use of these forests, although it may be locally significant for individual landowners. Logging is, however, an important secondary use of the Watershed, but as there is virtually no public access, the recreation potential of this area has yet to be realised. In contrast, much private forest land, when accessible to the public, is utilized for a wide

67

FIGURE 19,2
Varied scenes in the urban forest.

FIGURE 20,2
Variety enhanced by forest–water contrasts.

range of outdoor recreation. Indeed it is probable that alienated lands provide as many recreation man-days as do the regional parks, although figures are not available.

The forests are predominantly used for low intensity traditional forms of outdoor recreation such as hiking, picnicing, swimming, horse-back riding, fishing and general pleasure driving. More intensive uses occur along forested waterfront, lakeside and streamside areas (Figures 21,2 and 22,2). Where forest lands have been aquired for public park and recreation purposes by the Capital Regional District, as for example, North Hill, Bear Hill, Elk/Beaver Lakes, Durrance Lake and Mill Hill parks, (Figure 25,2), the basic philosophy governing management is "environmental" in aim, geared to protecting 'natural' environments for extensive, mainly appreciative forms of recreation. Development is limited to the provision of walking trails and parking lots at park boun-daries with the exception of Elk and Beaver lakes which contain picnic and beach area facilities. It must be recognized, however, that to some extent this situation results from financial-political conditions inherent in the division of funds and responsibilities between various government levels and agencies, and not only from basic preservationist thinking. Funds were available to acquire forest land in a "land-bank" acquisition programme; but since that acquisition has been completed, little money has been available for or channeled into management and development of the acquired areas.

The responsibility for developing 'intensive' facilities in forest parks has fallen upon the municipalities and the Provincial Parks Branch in a few of the parks that they control (Figure 23,2). But even here, development means little more than the traditional and familiar picnic tables, car parks, toilet facilities and beach areas. These are often placed in forest settings which, it is hoped, will withstand the inevitable heavy pressures and still retain their ecological values for aesthetic ap-

FIGURE 21,2
Such viewpoints are popular.

FIGURE 22,2
Any luck?

preciation and nature interpretation programmes (Figure 24,2). That hope, it is increasingly clear, is a vain one; and the whole matter of the validity of a preservationist, environmentalist and low-intensity use philosophy as the dominant rationale in the management of an urban forest is also open to serious question.

BASIC PREMISES

The purpose of this chapter is to stimulate ideas but not specify particular management techniques for the further humanising of the Victorian urban forest. But before putting forward suggestions for reappraisal of the present system, some basic observations and premises should be stated and briefly elaborated. These premises stem not only from the author's personal convictions gained from work in this and other urban forests. They also represent a concensus of opinion emerging in many urbanised areas, both American and European, that a reorientation, rethinking and intensification of forest management is necessary if the urban forest is to survive and play a dynamic role in social development.

1.   Forest management in urban and peri-urban areas should be 'people-oriented' and not 'tree-oriented'. The aims of managing forest land for urban, recreational, amenity and industrial uses are to create and maintain optimum benefits for, and values determined by, society: specifically, the urban society which bears the ultimate responsiblity for deciding what kind of environment it wishes to live in. A rigid adherence to previously accepted environment protection principles may therefore in some cases have to be abandoned. For whilst objective inventories of resource potentials or ecological parameters are possible, actual management decisions are conditioned by social objectives, not ecological imperatives or economic dictates. Social values and aspirations cannot be set on 'inherent values' or 'inherent capabilities' of land or resources.

FIGURE 23,2
Signs of intensive development?

FIGURE 24,2
Watching the salmon, Goldstream Provincial Park.

2.   Therefore, a prerequisite to the planned humanising of the Victoria urban forest is knowledge of peoples preferences and desires in addition to knowledge of physical/biological resource characteristics and capabilities. There is little doubt, for instance, about what 'minority' groups are seeking, be they bird-watchers, wilderness back-packers or white-water canoeists. But who knows what the urban-based public expects and hopes for from forest environments and experiences in the areas visited at weekends? If the European evidence is any indication, then much forest management (or lack of it) around North American cities has in the past appeared to be designed to turn people away rather than attract them.

3.   There is a need to widen the range of recreation and other social benefits of forest areas through increased provision of urban-type or intensive recreation facilities in more tolerant forest environments. Many people use 'wildland' areas within the urban forest for needs which would be better met in forests managed for heavier use. It may be possible to relieve pressures on intolerant and fragile areas through the development of alternative recreational facilities. Additionally, there may be a large latent demand for such intensive facilities, which is presently not satisfied. In the forest parks close to or within the Victoria urban area, the emphasis on relatively extensive or "basic unorganised outdoor recreation" and minimal facility development may well deny realisation of optimum social benefits and also preclude a much wider use of those parks. We have yet to face up to the full implications of the fact that whilst we are dealing with forests, we are also planning the environment of an urban society. The creation of diversity in the range of recreation opportunities available should therefore be a major aim of managing the urban forest.

4.   Similarly, the preservation of diverse and high quality or unique biological environments within the urban forest is also an integral aim of management, albeit one of the most difficult in the face of rising population pressures. This will entail active management and not merely

74

passive preservation, especially in seral and fragile areas, if their values are to be protected and maintained. It is also necessary to more clearly define the quality characteristics of these areas in light of the increasing conflicts with other potential users or uses. Means whereby it may be possible to create quality environments rather than just preserve those already existing should also be examined.

5.   Over much of the urban forest, management should be determined primarily by considerations of aesthetics, social and environmental amenity and recreation. Fibre forestry may be a by-product of management but not a prime aim. Additionally, legislation should ensure that people have reasonable access to public and private forests for recreational purposes. Considerations of safety and health, environmental protection and indivi-dual or industrial rights will vary from place to place and may influence particular access decisions.

6.   Historical precedent or tradition, profit maximization or legislative and administrative inertia are not valid reasons for maintaining the essentials of public land use as they are. Even private land owners have a moral ob-ligation in their use of land. Therefore, whilst acknowledging the tangible economic or environmental benefits of some present forms of forest use and tenure, urban forest management should examine those benefits closely with respect to intangible social values and the changing resource base and needs of the region.

7.   Forest environments of the Victoria region can provide a much wider range of benefits and values than they do now. But deliberate and long-term planning are necessary to realise this potential, and in the absence of such planning, benefits from future forest environments may well decrease. The total range of possible benefits for the whole region can best be deter-mined and planned for if the geographically separated, ecologically con-trasted and recreationally varied forest areas are designated and managed as parts of an integrated urban forest system.

THE FUTURE OF THE URBAN FOREST

Planning Approaches

The need to reappraise the roles and future management of Victoria's urban forest demands a co-ordinated planning framework over and above that provided by current planning categories and thinking. Although conceived as part of an overall regional plan by the Capital Regional District,[12] present forest planning and land use categories tend to perpetuate thinking along compartmentalised lines and underplay the complementary relationships between the separate forest areas. Designation of all private and public forested lands as an Urban Forest Park would appear to have considerable merit (Figure 25,2). The major components of the Urban Forest-Park would be:

1. All existing forested and partly-forested regional, provincial, municipal and federal parks within the metropolitan area;

2. All forested areas designated as Major Park Areas in the Victoria Metropolitan Area Regional Plan (permitted uses in that Plan - agriculture, water supply and recreation);

3. All areas designated as Upland Areas in the Regional Plan, including the Goldstream watershed (permitted uses - water supply, forestry, agriculture, recreation and low density residential);

4. That area of the Highlands designated as Potential Urban Area in the Regional Plan;

5. Parts of designated Rural Residential Areas in the Regional Plan, especially in the "western community" of Colwood, Langford and Metchosin, and the northern Saanich peninsula (permitted uses - low density residential, agriculture, and recreational, institutional and commercial development for local populations).

The bulk of the Urban Forest-Park therefore occupies the forested uplands to the west of the major built-up and agricultural areas. But also included are small, very humanised, yet still valuable wooded landscapes within established urban areas. These play a critical role in urban envir-

76

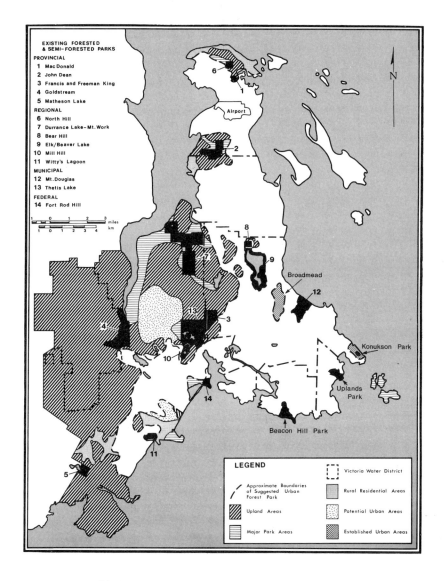

EXISTING FORESTED
& SEMI-FORESTED PARKS

PROVINCIAL
1 Mac Donald
2 John Dean
3 Francis and Freeman King
4 Goldstream
5 Matheson Lake

REGIONAL
6 North Hill
7 Durrance Lake-Mt. Work
8 Bear Hill
9 Elk/Beaver Lake
10 Mill Hill
11 Witty's Lagoon

MUNICIPAL
12 Mt.Douglas
13 Thetis Lake

FEDERAL
14 Fort Rod Hill

Airport

Broadmead

Konukson Park

Uplands Park

Beacon Hill Park

LEGEND

Approximate Boundaries
of Suggested Urban
Forest Park

Upland Areas

Major Park Areas

Victoria Water District

Rural Residential Areas

Potential Urban Areas

Established Urban Areas

FIGURE 25,2   The components of Victoria's urban forest park.

onments and society, complimenting the wilder forest landscapes of the peri-urban setting. Uplands Park, a Garry oak-spring flower preserve in Oak Bay; Konukson Park and its surroundings, a scenic coastal fir-arbutus preserve on Ten-Mile Point; Broadmead, a forested high quality residential development area in Saanich, and parts of the forested-agricultural residential landscape mosaic of Metchosin and Langford, fall within this category.

The major value of a Forest Urban-Park designation would be to recognise that in terms of function (serving the diverse recreation, residential and amenity needs of the region's population) the geographically separated forest areas are parts of an integrated system and should be managed as such. The Forest Urban-Park concept overlaps with that of the "Greenbelt", differing from it in excluding dominantly agricultural areas, but including rural residential and parts of established urban areas. Together, the two concepts form a basis for future urban and peri-urban planning. Involved in the management of the Park is the maintenance and improvement of forest landscapes and environments for ecological viability, aesthetic quality, wildlife habitat and both intensive and extensive forest and water-based recreation. Also integral to the concept is the environmentally-conscious development of residential, industrial and 'urban playground' recreation facilities which can be integrated into forest landscapes given sufficient forethought, ecological knowledge and planning committment.

One outstanding feature of European forestry is the manner in which forests adjacent to many cities are managed as sources of diversified public services rather than as purely production forests. A number of studies have shown that values placed on protective, amenity and recreation functions of these forests now largely exceed values from timber yield and traditional uses.[13] An Urban Forest-Park in the Greater Victoria region would closely

parallel examples such as Oslo, where the city owns 38,000 acres of production and recreation forest; Zurich, where 24 percent of the city is
forested; Copenhagen, which contains the 6,500 acre Jaegersborg Forest,
used intensively as an urban playground and production forest, and which
epitomises the appreciation of rural and urban values in forest management;
and forests around the larger cities of Czechoslovakia, managed for joint
recreation/timber use, and which are linked to the cities by well developed
public transport systems.[14]   A slightly less close analogy would be with
some of the National Parks in Great Britain.

A 'resort' component therefore enters into much European urban
forestry to supplement the 'resource' component of management.   One
British forester has argued that:

> Forests are needed within easy reach of the
> conurbations...not to produce timber or make
> money, not even to preserve or prettify the
> environments; but to attract and absorb the
> people who are ravaged by urban society.[15]

To such people, it is suggested, nature trails, "multiple use", even picnic-
sites, are meaningless. It is thus necessary to go beyond the provision of
the familiar camp site, picnic-area and carpark facilities and provide amen-
ities specifically to attract urban dwellers. Forest - placed urban facilities
such as hotels, race-tracks, game parks and dance halls, would act as at-
tractions and perhaps also as stepping-stones to more general forest - based
recreation.

It is not the intention of this author to wholeheartedly endorse such
a philosophy. One could question to what extent Victorians are 'ravaged'
by their urban environment; and nature trails, picnic-sites and forest reserves obviously are meaningful to much of the local population. But there
is no doubt that Victoria's urban forests can be used much more intensively,
whilst still retaining and increasing their many qualities; and there can

be no doubt also, that the demand for changed and more intensive management directly related to the needs of urban dwellers, will increase in the future. Some environments, especially the Douglas-fir-salal and grand fir-swordfern types are relatively resistant to visual and ecological change, and can absorb, or if necessary, hide such intensive developments. Even the more fragile oak-rock outcrop-bulb meadow, arbutus and arbutus-fir areas, because of their complex vegetation/landform pattern and textures, can absorb visual change to some degree if problems of ecological vulnerability can be reduced.

Purists may shudder at the thought of 'resort' type developments in, for example, Thetis Lake, Elk/Beaver Lake or Mount Douglas parks. Yet we are part-way there now in a piecemeal, rather uncomfortable fashion. Why cannot there be another, improved, Stanley Park (Vancouver's prime semi-forested scenic and pleasure ground) in the Victoria Urban-Forest Park? Why not open up some of the area's forested regional parks, such as Durrance Lake - Mount Work, to camping? Why should the bulk of the urban forest remain as environmental reserves for the bushwalker or nature-lover? We have a forest cover that is extensive and varied enough to accomodate, with planning, all the future needs of its urban users.

The value of looking at the European experience in urban forestry is not that it necessarily provides us with specific techniques of forest management which could be applied in the local context; many are clearly inappropriate, others may be worth consideration. But what is instructive and thought provoking is the diversity and intensity of management approaches and thinking that can be examined to retain high quality forest environments. It is true that a low ratio of forest resources to population and recreation demands has often encouraged the intensity of use so characteristic of European urban forests. But we should recognise that with increasing urbanisation of the Greater Victoria area, we can only look forward to a similar situation,

80

and that our present high ratio of urban forest areas to urban population numbers and demands will soon be a thing of the past.

Much land within Victoria's urban forests is privately owned, and some degree of public control over or direction of use of this land is essential in the context of total regional forest-use planning. There is widespread and undoubtedly deep-felt opposition to such a 'socialist' philosophy, but in European countries such as England, Germany and Denmark, controls of this nature are accepted and apply to almost all private forest land. Attention is also being given to the plight of private woodlots and woodlot owners in many urbanised areas of eastern North America.[16] As the dominant planning authority for the region, the Capital Regional Board, in conjunction with provincial and municipal authorities, should examine the ways in which existing land legislation, zoning and taxation could be ammended to allow more 'public-oriented' or 'socially-oriented' management of private forest land, and adequate compensation for economic losses or costs incurred by landowners be-cause of such controls. The possibilities include, for instance:

1. The use of tree or forest "ordinances" under which landowners and real estate developers would have to apply for permits before fel-ling any trees, and may also have to apply particular silvicultural techniques to maintain or improve scenic, amenity or ecological values;

2. The opening of all private forest land to public access;

3. Negotiated agreements under which local government or planning authorities could install recreation facilities such as walking trails, parking areas, picnic areas, playgrounds and lookouts on private land designated as recreational areas or heavily used for recreation purposes.[17]

The question of public access to forest land is crucial to the future of a significant area of the Victorian urban forest — the 15,000 acre Goldstream watershed — part of the larger Victoria Water District catch-ment. In common with 70 percent of municipal watersheds in the Pacific northwest states,[18] and all of them in coastal British Columbia, the Goldstream watershed is virtually closed to the public.[19] General access to the area for water or forest-based recreation is thought by the District to be incompatible with the production of safe potable water. And although the Goldstream section of the watershed does not now contribute to municipal drinking supplies, it is intended to be fully integrated into the supply system by 1983. That system contains only simple chlorination treatment; fears of human contamination in the absence of further treatment, and the economic costs and construction problems of supplying such treatment have so far been effective barriers to increased public access. Other arguments used against pressures for recreational use have been that public access would provide recreation benefits for only a minority at a cost to the majority, and that there is enough land for recreation outside of the watershed.

These pressures for public access to the watershed have not been broadly-based or intensive, and have frequently been based on misinformation. But they could intensify, particularly if parts of another extensive forest area — the Highlands — are given over to residential development. Additionally, by taking an exclusionary view point with regard to watershed use, the Water District is imposing a professional value - judgement on the public; the validity of this as an adequate criterion for public decision-making is questionable. There has been minimal public participation in the little debate that has taken place, and as in other, similar, situations, much of the information necessary for decision-making is unavailable, unasked for or not freely exchanged between the different parties involved.[20]

But the potential recreation and amenity, as well as water and timber values of the watershed area should be acknowledged.[21] Proximity to a large and expanding urban area of such a varied and often aesthetically attractive forest and lake region is a significant factor.

82

Future energy and gasoline conditions may well force a decrease in peoples' recreational mobility, and the value of high quality recreation areas close to city centres will be increased. It would therefore seem appropriate to revive and review arguments forwarded by the Capital Regional Board in 1969 for the inclusion of the Goldstream Watershed in the regional parks system,[22] and therefore in the Urban Forest-Park.

Some Research Suggestions

The foregoing thoughts on the broader aspects of planning for the future of the urban forest certainly do not cover all the options that are open. Intended hopefully to stimulate ideas, it is also hoped that they are specific enough to provide some input into future management debates. The following brief observations suggest some of the more concrete research and management avenues that could be followed to initiate a socially relevant and ecologically sound approach to the management of the urban forest.

Existing data on land capabilities and characteristics are as yet preliminary and generalised. For example, except where associated with water bodies, scenic viewpoints or "unique and attractive vegetation", the urban forest is classified as greenbelt of only moderate recreation significance.[23] Use-intensity categories have been applied only to waterside corridors (Figure 26,2), but refined capability and recreation categories are also needed for the more extensive forest areas away from the immediate vicinity of lakes, streams and ocean (Figure 27,2). Such information is necessary to supplement and, maybe, modify the existing framework of planning categories, and to provide a partial basis for optimizing use intensity and quality, protecting landscape and ecological diversity and reducing user conflicts. Priority in this data collection should initially be given to private forest lands rather than public lands for which use controls are relatively well established.

83

FIGURE 26,2
Recreation values of such areas are obvious.

FIGURE 27,2
What values are offered by these areas?

Land capability, however, is only one of the factors influencing
the value and ultimate use of any piece of forest. Recent environmental
quality ranking systems that have been developed to estimate the qualitative values of many natural and cultural landscapes indicate intriguing
applicability to the urban forest scene. In particular, the approaches
used by Fines for rural and urban landscapes,[24] Leopold and Marchand
for riverscapes[25] and Litton for forests[26] could be modified to produce
quality ranking systems for the urban forest. These systems would be
based upon a number of objectively and subjectively determined ecological, aesthetic, amenity and physical attributes of the forestscape, and
could be used to evaluate the suitability of both single-purpose and composite potential uses. Individual areas could be placed on a rating
hierarchy or scale relative to all other areas of the urban forest. Similarly, biological or recreational "uniqueness ratios" could be developed
for all areas relative to others, and these ratios could be of help in
evaluating 'trade-off' situations where particular use or management alternatives have to be balanced.

Much of the quality of Victoria's urban forest stems from its varied
and dynamic seral characteristics. It is in the ecological nature of these
communities to change relatively rapidly. Management that confines itself to preservation and protection is therefore inadequate: it must em-
ploy silvicultural techniques to retain and create desired and agreed upon
values of these dynamic communities. This crucial aspect of management
has yet to be developed, and the forest parks in the region are managed
as if they were static entities. There is a large field of research open to
examine the ways in which ecosystem characteristics, composition, quality
and diversity can be deliberately controlled by various techniques of
vegetation and site manipulation.

Some of the areas that appear fruitful include:

1. Detailed studies of the dynamics of seral forest types, particularly the factors influencing the successional relationships of Arbutus and Garry oak to conifers;

2. Related studies of successional changes in understory and ground cover vegetation and flora associated with changes in tree density and species mix;

3. An assessment of the feasibility of using such techniques as selective or shelter felling, ground scarification, control burning or planting to retain or create desired tree species mixes, aesthetic appearance and stocking densities in the different forest-types;

4. An assessment of techniques that could retain or increase vernal floristic diversity, especially in fragile or disturbed areas;

5. Studies of ways in which significant, unique or fragil areas such as wetlands, vernal meadows, moss-lichen rock outcrops or coastal headland vegetation, can be protected from heavy use by careful site or facility design, rather than by denying public access;

6. Related to 5, detailed examination of techniques such as paved footpaths, picnic and camp-site areas, fences, guide and control systems or other methods whereby capital outlay, facility design and management could supplement and increase carrying capacity of recreation areas in general and allow greater use of those areas without degrading the resource.

7. An examination of the problems, potentials and advantages of retaining at least partial tree and understory cover in suburban and high density residential areas, if only to save homeowners the long, tedious and costly business of attempting to regrow their own privacy screen!

In addition to these ecologically or silviculturally oriented research needs there are also unexplored avenues of research into peoples' preferences for and perceptions of various forest landscapes and recreation opportunities. What do various forest environments mean to different groups of people? How do these groups perceive management alternatives affecting the type and quality of those environments? How do they perceive alternative forest-based recreational facility developments, and how com-

patible are the goals and expectations of various users?[27]   These and related questions and the answers to them are crucial to future management
decisions in the urban forest.

CONCLUSION

The trees and tree-dominated landscapes within and around Victoria
are an even more characteristic and valuable resource than are the horticultural delights of this "garden city". Few other cities can lay claim to such
diversity, aesthetic appeal, amenity value and recreation potential. But
in spite of those qualities, and because of present planning and management
shortcomings, Victoria's forests have yet to reach their full potential as an
urban forest. Without re-oriented management, high quality will be difficult to maintain in the face of future urban trends and increased pressures.
Such management, in its objectives and techniques, will have to refute
many existing ideas and doctrines of what forest management should be.
It must firmly grasp the fact that in planning for ecological viability, our
starting point is the creation of a high quality, satisfying living environment for an increasingly urbanised population.

Much of the urban forest now lying outside the urban area, will,
if it remains, be enclosed by that area in the year 2000. Changing forest
characteristics associated with that transition, and changes also inherent in
the dynamic seral nature of much of the forest, emphasise that this humanised landscape must be actively managed to retain or attain agreed-upon
values. Parodoxically, some of that management will not be directly involved with forest land, but must ensure more intensive use of existing
open space, and the development of outdoor recreation environments,
facilities and experiences within the city. By so doing, it can provide
alternatives to forest or wild-land recreation for many purposes, and may
lessen the need for 'escape' from the built-up environment. For no matter
how comprehensive urban forest management becomes, it will be severely

87

strained by a mass weekend exodus of city dwellers seeking in the forest environment something denied them closer to home. And preserved fragile, unique or critical ecosystems, integral to the urban forest, will certainly disappear under such pressure.

But preservation of wilderness or 'natural' environments is not the major aim of urban forest management; neither is the conversion of all forest areas into urban pleasure grounds the dominant philosophy. Somewhere between these two extremes, yet with the full appreciation of our urban existence, we can plan not to "reduce the impact of man on the forest, but to increase the impact of the forest on man."[28] If we can move away from thinking of the city and the forest as two distinct entities; and if we can plan and use the forest as an integral part of the urban environment; then in humanising the urban forest we can take an essential step in developing a regional urban ecology.

REFERENCES

1.   Some of the ideas expressed in parts of this chapter were initially
      developed in a report entitled "Study and Review of the
      Socio-Economic Use of Forest Land in Urban and Peri-
      Urban Areas" prepared by the writer for Environment
      Canada, Pacific Forest Research Centre, under Research
      Contract PC - 31 - 194X(1974). It should be emphasised,
      however, that those ideas do not reflect either policies
      formulated or opinions held by Environment Canada.

2.   The title and general theme of this chaper were decided before the
      writer became aware of Dubos' paper "Humanising the
      Earth" in Science, 174 (1973), pp. 769-772. Dubos ar-
      gues that in the global shaping of nature by culture, man
      has, in many instances, actually improved on nature by
      bringing out potentials and diversities unexpressed in the
      state of wilderness. He points out that active management
      is necessary to maintain these high quality humanised en-
      vironments. Humanising the earth entails not only the
      creation of man-made nature but also a complementary
      preservation of wilderness "where man can experience
      mysteries transcending his daily life," (p. 772). These
      are also the themes which are here examined in the nar-
      rower confines of the urban forest.

3.   Ibid., p. 769

4.   SMITH, D.M. "Adopting Forestry to Megalopolitan Southern New
      England," Journal of Forestry, 67, No. 6 (1969), pp.
      372-377.

5.   See, for example, HOPKINS, W.S. "Are Foresters Adequately
      contributing to the Solution of America's Critical Social
      Problem?" Journal of Forestry, 68, No. 1 (1970), pp.
      17-21, and HIGGS, K.G. "Changing Forestry Environ-
      ment; Resources Management in the Urban Colossus,"
      Forestry Chronicle, 44 (1968), pp. 24-25.

6.   JORGENSEN, E. "Urban Forestry in Canada," Faculty of Foresty,
      University of Toronto, (1970), pp. 1-15.

7.   RUCKELSHAUS, W.D. "Forestry in an Urbanised Society," Journal
      of Forestry, 69, No. 10 (1971), pp. 712-714.

8.   See FORWARD, C.N. Chapter 1 in this volume.

9.   For fuller descriptions of forest vegetation in the area, see ROEMER,
        H.L. Forest Vegetation and Environments on the Saanich
        Peninsula, Vancouver Island. Ph.D. Thesis, University of
        Victoria, B.C. (1972); STANLEY-JONES, C.V. and
        BENSON, W.A. (eds) An Inventory of Land Resources and
        Resource Potentials in the Capital Regional District,
        Capital Regional District, Victoria, B.C. (1973); and
        EIS, S. and OSWALD, E.T. The Highland Landscape,
        Environment Canada, Forestry Service, Victoria, B.C.
        (1975).

10.   COOK, W.L. "An Evaluation of the Aesthetic Quality of Forest
        Trees," Journal of Leisure Research, 4, No. 4 (1972),
        pp. 293-302.

11.   DUBOS, R. op. cit., p. 772.

12.   Capital Regional District, Official Regional Plan, Victoria Met-
        ropolitan Area, Schedule A. Victoria, B.C. (1975).

13.   For example, PRODAN, M. "Evaluation of Forests for Multiple
        Use," Schriftenreihe Forstlische Abteilung, 4 (1965), pp.
        34-50.

14.   See BARTA, C. (ed.) Czechoslovak Forestry. Ministry of Agriculture
        and Forestry, State Agriculture Publishing House, Prague,
        (1966), and HOLSCHER, C.E. "City Forests of Europe,"
        Natural History LXXXII, No. 9 (1973), pp. 52-54.

15.   RICHARDSON, D. "The End of Forestry in Britain," Advancement
        of Science, 27 (1970), pp. 153-163.

16.   See, amongst others, DAVIS, K.P. "Land: The Common Denomin-
        ator in Forest Use Management, Emphasis on Urban Relation-
        ships," Journal of Forestry, 68, No. 8 (1970), pp. 628-
        631, and HAMILTON, L.S. "Private Woodlands, the
        Suburban Forest and Aesthetic Timber Harvesting," Forestry
        Chronicle, 44, No. 2 (1966), pp. 162-166.

17.   A recent move in this direction was heralded by the passage of The
        Recreational Land Green Belt Encouragement Act by the
        Provincial Legislature in 1975. Under this act a land-
        owner can apply for his land to be designated as approved

recreational land, enter into a management agreement with the government for the land to be used for public recreation activities, and for this receive reimbursement of property taxes.

It is interesting to note that a survey of the opinions of 405 resources, environmental and ecological experts has recently suggested that by 1985 economic incentives will be widely available to private landowners who open parts or all of their land for public recreation. See SHAFER, E.L., MOELLER, G.H. AND GETTY, R.E. Future Leisure Environments, USDA Forest Service, Research Paper NE-301, N.E. Forest Experimental Station, Upper Darby, Pa., (1974).

18.   BAUMANN, D.D. "Perception and Public Policy in the Recreational Use of Domestic Water Supply Reservoirs," Water Resources Research, 5, No. 3 (1969), pp. 543-554.

19.   The exception is the issuance of permits for a limited number (less than 100) of medically-passed deer hunters each hunting season.

20.   KASPERSON, R.D. "Political Behaviour and the Decision making Process in the Allocation of Water Resources between Recreational and Municipal Use," Natural Resources Journal, 9, No. 2 (1969), pp. 176-211.

21.   The Water District does, however, show some appreciation of the recreational value of the Goldstream Watershed. See HOMER-DIXON, D. "The Victoria Watershed: Conflict Resolved," a paper presented to the "Symposium on Practical Forest Watershed Management," sponsored jointly by the University of British Columbia Extension Department and the Association of British Columbia Professional Foresters, Parksville, B.C., November 1973.

22.   Capital Regional Planning Board, Regional Parks, Victoria, B.C. August 1969.

23.   BENN, D.R. "Recreation," in STANLEY-JONES, C.V. and BENSON, W.A. op. cit., pp. 159-171.

24.   FINES, K.D. "Landscape Evaluation: A Research Project in East Sussex," Regional Studies, 2 (1968), pp. 41-55.

25.   LEOPOLD, C.B. and MARCHAND, M.D. "On the Quantitative
      Inventory of the Riverscape," Water Resources, 4, No.
      4 (1968), pp. 709-717.

26.   LITTON, R.B. Forest Landscape Description and Inventories –
      A Basis for Land Planning and Design. USDA Forest
      Service, Research Paper PSW-49. Pacific S.W. Forest
      and Range Experimental Station, Berkeley, California,
      (1968). See also, LITTON, R.B. "Visual Vulnurability
      of Forest Landscapes," Journal of Forestry, 72, No. 7
      (1974), pp. 392-397.

27.   For discussions of such approaches, see MOELLER, G.H.,
      MacLACHLAN, R. and MORRISON, D.A. "Measuring
      Perception of Elements in Outdoor Environments, USDA
      Forest Service, Research Paper NE-289, NE. Forest
      Experiment Station, Upper Darby, Pa. (1974); SHAFER,
      E.L. "Perception of Natural Environments," Environ-
      ment and Behaviour, I (1969), pp. 71-82; and SHAFER,
      E.L., HAMILTON, J.F. and SCHMIDT, E.A. "Natural
      Landscape Preferences: A Predictive Model," Journal
      of Leisure Research, 1 (1968), pp. 1-19.

28.   SPURR, S.H. and ARNOLD, R.K. "The Forester's Role in Today's
      Social and Economic Changes," Journal of Forestry, 69,
      No. 11 (1971), pp. 795-799.

CHAPTER 3

TOPOCLIMATIC PATTERNS OF NOCTURNAL TEMPERATURE ON THE SAANICH PENINSULA

Stanton E. Tuller

University of Victoria

and

Rodney Chilton

British Columbia Environment and
Land Use Committee Secretariat

One of the tenets of modern climatology holds that the variation
in the nature of the earth's surface from place to place is a prime factor
in creating local differences in climate. Grigoryev and Budyko, for
instance, state that there are only two ultimate controls of climate:

1. The amount of available solar energy incident on the top of the
    earth's atmosphere and,

2. The nature of the earth's surface.[1]

This surface–atmosphere interaction has important implications
for man's utilization of the earth's surface. The local climate coinciding
with a specific surface type will offer certain constraints or advantages
for any particular land use. On the other hand, man's use of the surface implies at least some degree of modification of the surface character
which will, in turn, alter the microclimate of the affected area.

The recognition of the important interaction between man's activities, the nature of the earth's surface and local climate has dictated
that much research in climatology be directed toward the investigation
of the relationship between climate and surface type. These studies
have proceeded on a variety of scales. Those most often considered are

the micro and meso-scales. Microclimatology studies the climatic characteristics and causal processes found over precisely defined surface types. Slight variations in slope angle or aspect, soil type, surface moisture, or vegetation type are enough to create a different micro-climate.

Meso-scale climatology focuses on the climatic effects created over more broadly defined segments of the landscape. Common comparisons include those between land and water surfaces, different slope aspects, concave and convex landforms, and rural and urban environments. Minor variations within the broader divisions are disregarded.

Meso-scale studies have often been designated "topoclimatology." The term derives from the influence of topography on the local climate. Topography is used in its broader sense referring to "surface features of an area, including not only landforms but all objects and aspects both of natural or human origin."[2] Topoclimatology offers the advantage of combining enough spatial resolution so that practically important differences in climate may be delimited with an economy of observational effort which makes a study feasible within the constraints of limited instrumentation and manpower.

The Saanich Peninsula offers an ideal laboratory for the study of topoclimate. Its varied landscape includes a complexity of hills and valleys, urban and rural environments, various sizes of inland water bodies, the juxtaposition of land and sea surfaces, and a variety of land uses. Much of the peninsula is intensively used for agriculture, live-stock raising or suburban residence so that local differences in climate will have an immmediate impact on man and his activities.

The present study looks at the topoclimatic spatial variation in one important element, nighttime air temperature. Nocturnal air temperature is important in a number of practical applications including plant growth, frost risk, number of diurnal freeze thaw cycles, home heating requirements and the interruption of transportation because of the formation of ground fog.   This paper will present the pattern of nocturnal air temp- erature as found in the central portion of the Saanich Peninsula on clear nights with light winds when areal differences reach their maximum. The relation between the temperature pattern and topography will then be discussed followed by some examples of the practical consequences of this pattern for horticulture, home heating and fog frequency. Em- phasis is on the effects of the land–sea distribution, convex and concave landforms, and rural and urban environments.

METHODOLOGY

The study area comprises the central portion of the Saanich Pen- insula running from approximately Elk Lake on the south to Victoria In- ternational Airport on the north (Figure 1,3).   Elevations within the area range from sea level to just over 1000 feet (305 m) at the top of Mt. Newton.   Landuse within the area consists of a complex intermix- ture of woodland, pasture, horticulture and residential development.[3]

Nighttime temperature data was collected using the mobile tra- verse technique.   A thermistor thermometer was mounted in the front of an automobile at a height of 28 inches (71 cm) above the road sur- face.   The thermistor bead was protected by a radiation shield and lo- cated far enough in front of the automobile to minimize any effects of engine heat.   The instrument has a response time of approximately one second.

A number of transect routes were established covering the major topographic features of the Peninsula.   Because of the total length, all routes could not be conveniently covered in a single night's observation. The temperature pattern portrayed in Figure 2,3 is the composite gain- ed from over twenty nights of observation spanning the period from the

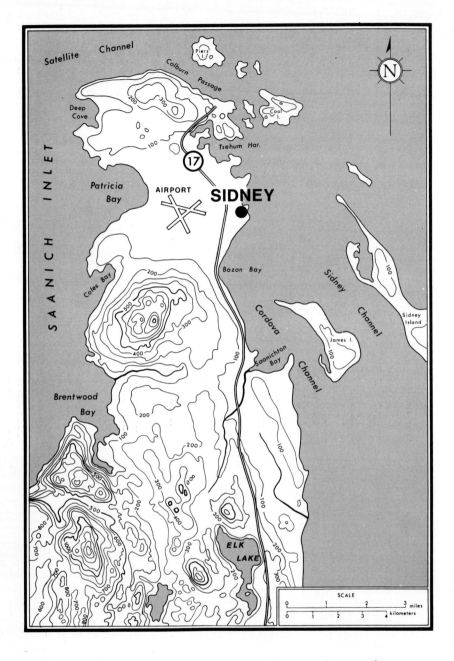

FIGURE 1,3   Location and relief map, Saanich Peninsula.

96

spring of 1973 to the spring of 1975. Temperature readings were re-corded every one-tenth of a mile (.16 km), or whenever a noticeable change in temperature occurred, by means of a tape recorder. More detailed analysis makes use of ten nights of data taken during the spring season when the importance of nocturnal temperature is significant for agriculture, and the proper combination of clear, radiation nights re-inforcing seasonally cool temperatures makes the risk of exceeding criti-cal low temperature limits rather high. Forty-one stations that repre-sent a variety of topographic situations were chosen for this analysis.

Observations were made between mid-night and 6 A.M. Nights selected for mobile temperature runs were those with clear skies and light winds when topographically determined differences in temperature reached a maximum. Clear skies allow the maximum rate of radiational cooling. Under these conditions the air temperature immediately over any surface is largely controlled by the availability of stored energy which can be released to warm the air and offset a part of the radient heat loss. In general open water, moist soil, dense pavement or base rock have a greater storage ability and warmer nighttime temperatures than do dry soil or vegetation.

The cold dense layer of air that forms next to the surface becomes deeper as the night progresses. Eventually this layer becomes unstable and tends to flow downhill and accumulate in areas of concave config-uration. This drainage of cold air occurs in discrete, periodic pulses. Thus, a second important control of nocturnal temperature pattern is the distribution of concave and convex landforms. Convex forms (hills) or slopes that allow easy air drainage have warmer temperatures. Con-cave forms (valley bottoms) are cold. The latter are aptly called frost pockets or frost hollows because of the much more frequent occurrence of ground frost.

97

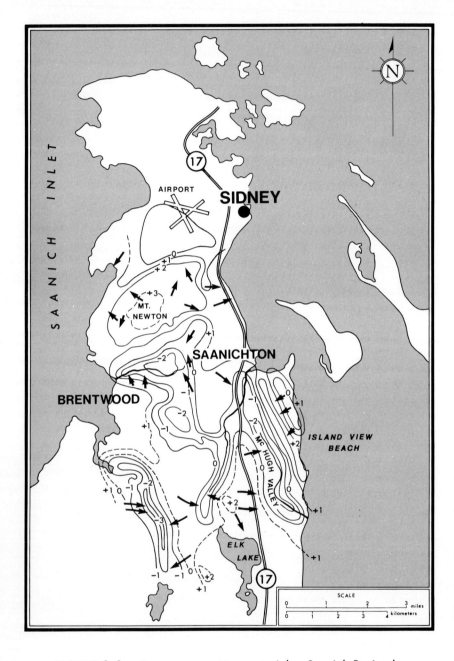

FIGURE 2,3   Temperature patterns at night, Saanich Peninsula.

98

During cloudy nights the cloud base will absorb and reradiate the longwave radiation given out by the surface. Only a small net cooling will occur and topographically determined differences in temperature will not develop to the degree found on clear nights. Strong winds will mix the air both vertically and horizontally producing a more homogeneous temperature pattern.

All study nights had cloud cover of three-tenths or less. Wind speeds measured at the airport averaged 5.25 m.p.h. (2.35 m/sec). Results, therefore, represent average clear night conditions. Cloudy nights, very windy nights and long term average values encompassing all weather types would show a more homogeneous temperature pattern. Calm, clear nights, however, would produce a more extreme differentiation than shown in this study.

The temperature readings taken at 28 inches with the automobile mounted thermistor were referenced to the screen temperature at Victoria International Airport. Recorded values were adjusted and results are presented as deviations from the airport temperature at standard screen height (48 inches, 122 cm).

Temperatures taken at standard screen height can be much warmer than the temperatures nearer the ground which are more likely to affect low growing crops and newly emergent seedlings. During the spring of 1975 a station was established in an open field near the centre of the study area. It was instrumented with Dobbie minimum thermometers placed:

1.   with a thermometer screen at standard screen height,

2.   at 28 inches (71 cm), and

3.   at 2.5 inches (6.4 cm).

An average of several clear nights showed the 28 inch temperature to be 0.8° C (1.5°F) colder than the standard screen height temperature. The 2.5 inch temperature was 3.0° C (5.5°F) colder. The latter was applied as a correction to temperatures at standard screen height when assessing the probability of exceeding certain critical temperatures for low growing crop plants and newly emergent seedlings.

NOCTURNAL TEMPERATURE PATTERN

Figure 2,3 shows the average pattern of clear night temperature distribution in the Central Saanich Peninsula presented as deviations from the temperature at Victoria International Airport. The microvariations in temperature have been smoothed over and the map presents the generalized meso-scale pattern. The temperature of each individual area is the result of a number of surface and atmospheric variables. Only the most important will be discussed here.

The major controlling factor of the clear night temperature pattern is the configuration of the land surface. The temperature isanomalies (lines of equal deviation) bear a strong resemblance to elevation contours. Areas of concave form (the valley bottoms) tend to accumulate cold air that drains from higher ground and have the lowest temperatures. This is a good example of the classic frost pocket situation so often described in the literature.[4] Temperature differences among valley areas themselves are largely a result of freedom of air drainage down the valley bottom, catchment area for cold air drainage and openness to moderating effects of the wind.

The coldest area in the Central Saanich Peninsula is the Tod Creek Valley which has an average temperature deviation of over 3° C and on especially favourable nights has been observed to be over 4°C cooler than the airport. The narrowness and depth of the valley allow a deep accumulation of cold air. The valley bottom is sheltered from the regional

winds in the area. This restricts the mixing of cold surface layers of
air with the warmer layers above. The north-south orientation of the
valley hinders the advection of warmer air from Saanich Inlet which lies
to the west.

The Hagan Creek Valley, Stelley's Cross Road - Wallace Drive
Lowland and the McHugh Valley are other areas of low temperature.
These areas, however, are more open to the wind, are shallower, have
a smaller catchment area for cold air and better outflow. Detailed
observations reveal an interesting contrast between the Hagan Creek
Valley and the Stelley's Cross Road-Wallace Drive Lowland (S W L).
The Hagan Creek Valley is about 0.5° C colder on the average than
S W L. The Hagan Creek Valley is deeper and tends to accumulate a
thicker layer of cold air during the night. Both detailed mobile transects
and a recording hygrothermograph in S W L show that after the layer of
cold air in S W L reaches a critical depth there is an outflow of cold air
northward following a minor stream valley into the Hagan Creek Valley.
This flow is accompanied by a definite warming in S W L followed by
slow cooling until the next outflow pulse. This and the more open exposure
of S W L gives it the most variable temperature pattern of the valley areas
and makes it the most difficult area in which to predict the value and tim-
ing of the nighttime minimum temperature.

The accumulation of cold air in the Hagan Creek Valley is aided
by a constriction near the mouth of the valley just to the east of West
Saanich Road. This effect is augmented somewhat by the road itself.
Observations show a deeper, colder layer of air to the east (inland) of
the constriction. The greater incidence of fog east of the constriction
helps to confirm the importance of even a small constriction acting as
an impediment to the free drainage of cold air.

Areas of convex form (hills and ridges) have the warmest clear
night temperatures. Cold air that forms near the surface via radiational

cooling can readily drain away and is replaced by warmer air from above. On most nights the warmest temperatures are found on the upper slopes of Mt. Newton. The ridge that separates Island View Beach from the McHugh Valley, although having a relative relief of little more than 100 feet (30 m) is over 4° C warmer than the bottom of the McHugh Valley to the west.

The advection of warmer air from the sea seems to have at least a small effect on the nocturnal temperature pattern. The average sea surface temperatures for Saanich Inlet are given in Table 1,3.[5]

TABLE I,3

MEAN MONTHLY SEA SURFACE TEMPERATURES FOR SAANICH INLET

Month	Temperature ($^\circ$C)	Month	Temperature ($^\circ$C)
January	5.7	July	19.0
February	7.0	August	15.5
March	8.5	September	15.0
April	11.5	October	13.0
May	11.5	November	5.8
June	14.5	December	5.2

Source: Herlinveaux[5]

Sea surface temperatures are usually warmer than clear nighttime air temperatures over the land during much of the year. During the spring observation nights emphasized in this study the sea surface temperature was an average of 6.3° C warmer than the airport screen temperature. Differences on individual nights ranged from 1.5° C to 11.8° C. Air advected from the water areas surrounding the Saanich Peninsula would be expected to produce higher temperatures. Indeed, a transect across the Peninsula in the vicinity of the airport where relief effects are not great shows that areas near the sea are 1° C to 2° C warmer than inland areas.

It is, however, impossible to separate the effects of the sea itself from those of slope or wind. Areas near the sea usually are sloped and allow free drainage of air out to sea. These areas are also better exposed to winds from off the water.

Personal observations of the patterns on the Saanich Peninsula seems to indicate that relief and surface configuration are more important than advection from the sea in determining nocturnal temperature. The area just north of Island View Beach offers an interesting example. This is a very shallow valley area and is the location of a frost pocket of about 1° C magnitude although it is located only 200 m. from the sea. Even the short distance to the sea is not enough to overcome the accumulation of cold air and this area is colder than the ridges both to the east and west.

A preliminary correlation analysis was conducted in order to further examine the relationship between difference from the airport temperature and various weather and location variables for 41 transect points over a period of eight spring days when airport temperatures were near or below freezing. The three variables with the highest simple correlation coefficients were: Distance from the nearest Valley Bottom, + .51; and Distance from the sea, - .44. All other variables tested had correlation coefficients lower than $\pm$ .20. This was not surprising for the weather variables such as wind speed and direction, dewpoint temperature and cloud cover because these were taken from measurements recorded at Victoria International Airport and were thus constant for each transect. The actual temperature, however, varied greatly from point to point. A variable such as wind speed would probably be much more significant if measured at each individual transect station.

A surprisingly low correlation (- .15), however, was found for the location variable "Distance Downwind from the Sea." This variable was a measure of the effective distance from the sea along the prevailing wind

103

direction as measured at the airport. It was thought that this would be a better measure of the influence of advection and the temperature moderating effect of the sea than the simple distance to the nearest shoreline. That this was not the case could be because:

1. wind direction measured at the airport may not be a good indication of local wind flow in other areas of the Peninsula,

2. other variables such as relief or slope vary systematically with distance from the sea and have a greater influence on temperature than does advection of warmer air from the sea,

3. the sea has some effect on air temperature for at least a small distance inland even without the aid of regional wind flow,

4. the influence of advection from the sea is truly of only minor importance, or

5. there were some undetected errors or noise in the data.

Further research is needed in order to determine the real reason for the low correlation between Distance Downwind from the Sea and measured air temperature and to sort out the true role of advection from the sea in determining nocturnal temperature patterns.

The preliminary correlation analysis indicates that it is indeed the surface configuration which is the dominating influence on nocturnal temperature patterns in the Saanich Peninsula. Other investigators have found generally good positive correlation between nocturnal temperatures and height above the valley floor in transects across single valleys. Harrison[6], for example, found correlation coefficients above .90 in the Low Weald of Kent. Lawrence[7], also in Britain, reported correlation coefficients of .66 and .75 for two individual May nights with pronounced radiational cooling. Hocevar and Martsolf[8] also found generally high correlation coefficients for the temperature-height relationship in the Nittany Valley of Pennsylvania, although values on individual nights ranged from .21 to .96. The actual relationship between temperature and elevation, however, is curvilinear with the vertical temperature gradient decreasing

with height.[9] The best correlations, therefore, are found over small
ranges of elevation.

It is not surprising that the correlation coefficient between
heights above the nearest valley floor and temperature found in the present
study is lower than those found in studies of single valleys. The 41
stations used in the analysis were selected to represent a wide range of
topographical situations. As such many were located in positions where
the nearest valley bottom was difficult to determine. Several stations
were located on slopes that ran directly to the sea so that no pooling of
cold air could occur. In these cases the shoreline was taken as being
the nearest valley bottom. As such, the overall correlation between
temperature and height above , or distance to, the nearest valley bottom
can not be expected to be as high as those where a detailed transect of
a single valley hold other factors more constant.

A stepwise multiple linear regression procedure indicated that the
equations:

$$\triangle T = -1.27 + .03 E + 2.28 D \tag{1}$$
$$\triangle T = -.26 + .03 E + 1.98 D - .51S - .11T \tag{2}$$

might be used to predict the temperature difference ($\triangle T$) between the airport and a particular location on the Saanich Peninsula during clear nights.
In the equations: E is the height above the nearest valley floor (or sea-
shore) (m), D is distance from the valley bottom (km), S is distance from
the sea (km), T is the airport temperature ($^{\circ}C$), and $\triangle T$ is in $^{\circ}C$. Equation
(1) has an R^2 value of .41 and a standard error of 1.25° C. Equation
(2) has an R^2 of .54 and a standard error of 1.12° C. Further research is
now being undertaken to improve the accuracy of predictive equations
such as those presented above. This work includes a more detailed observation of weather variables throughout the peninsula, more precise
measurement of location variables, an expanded network of mobile temp-

erature transects, the establishment of long term recording instruments in areas of particular topographical interest, and the investigation of other forms of predictive equations. It is hoped that a method can be developed which is accurate enough to allow the assessment of the temperature suitability of any particular area of the Peninsula for any specific land use or economic activity through reference to long-term airport climatic records. This would be an important planning tool for the individual landowner or regional planning authority and provide a much firmer climatic foundation on which land use decisions could be based.

In addition to the major factors discussed above, two other factors have been found to exert at least some influence on local air temperatures in the Saanich Peninsula.

Lakes, like the sea, have an effect in moderating nocturnal air temperatures in their immediate vicinity. The effect of lakes in moderating minimum temperatures and reducing the risk of frost has long been used to advantage and is well revealed in patterns of agricultrual land use throughout the world. The shores of Elk Lake, although not included in the primary study area, have been found to be about 0.5°C and in favourable circumstances up to 1.0° C warmer than similar areas further removed from the shoreline. Nurseries located just to the north of Elk Lake take advantage of this lake effect as well as the south slope of Bear Hill.

One of the most profound ways in which man is altering local climates is through the building of cities. The reduced reflection of solar radiation at the surface, greater heat storage in pavement and building materials, artificial heat generation, and reduction of evaporation from the impervious pavement surfaces are some of the important surface effects that give urban areas warmer nocturnal temperatures than rural areas. This zone of warmer temperatures in urban areas has been given the name "the urban heat island."[10]

The magnitude of the urban heat island increases directly with size of the city or more importantly with the density of the built-up area.[11] However, small urban areas, or even isolated shopping centres,[12] or airports,[13] have also been found to create nighttime heat islands.

The growth of residential subdivision in the Saanich Peninsula makes it an interesting area for studying the changes in local climate caused by the expansion of suburban housing and urban land use. One of the original purposes of the data collection program partially reported on in this study was to establish baseline temperature data with which the climatic effects of future changes in land use patterns (including urban expansion) could be measured.

Although on too small a scale to accurately portray on the map in Figure 2,3, some development of the heat island phenomenon was found in even the small settlements of the Central Saanich Peninsula. The transect routes passed through the village of Saanichton and on the outskirts of Brentwood Bay. Saanichton is a loose aglomeration of single family dwellings with scattered commercial establishments along the main road. West Saanich Road which passes through the eastern fringes of Brentwood Bay is lined with low commercial buildings and a small expanse of paved parking area.

Small heat islands averaging between 0.5° C and 1.0° C in Saanichton and a bit below 0.5° C in Brentwood Bay were evident from the observational data. Maximum values reached 2.0° C in Saanichton and 1.0° C in Brentwood Bay. Downtown Victoria, by comparison, is on the average about $2 - 3^\circ$ C warmer than nearby residential areas and $4 - 5^\circ$ C warmer than surrounding rural areas on clear nights.[14]

Observations did show, however, that the heat islands of these small communities were not enough to overcome the effects of surface configuration and distance from the sea. The temperatures in the eastern portion of Brentwood Bay were partially under the influence of the frost pocket of the Stelley's Cross Road – Wallace Drive Lowland. Temperatures were somewhat warmer than those at stations of similar relief and surface configuration but were always colder than nearby rural areas located on land with free draining slopes. Other investigators[15] have found that local hills, valleys and water bodies have a significant effect on local temperature even after urban effects are superimposed.

Local temperature variations should be an important consideration when planning the areas for further urban expansion. Considering only the climatic aspects it would be better to limit urban development to cold, frost pocket areas which are least suitable for high value fruit and vegetable crops. Here, the urban heat island effect may even be of some benefit in ameliorating nocturnal temperatures. The slopes should be preserved for agriculture.

Another important aspect that must be considered is the natural pattern of cold air drainage. Buildings, fences, road embankments or other obstructions should not be allowed to become cold air dams and thus, increase the extent and severity of frost pockets. Fonda[16] et al have documented the frost pocket created by a freeway embankment in Bellingham, Washington. Hawke[17] commented on a similar effect of a railway embankment in Britain. One of the authors of the present study during field work in the Kamloops area, observed one case which illustrates the practical consequences of cold air damming. In an area east of Kamloops the embankment created by the building of the Trans-Canada Highway impeded the natural drainage of cold air towards the South Thompson River. One farmer found that apple production in the portion of his orchard adjacent to the highway was severely curtailed by the increase in frosts and low temperatures resulting from cold air pooling behind the highway embankment.

APPLICATIONS

Climatic data are of little use unless applied to some practical situation. Since climate literally effects everything that occurs on the surface of the earth the number of possible applications of climatic data is virtually limitless. In order to illustrate some of the ways in which data on nocturnal temperature can be applied to practical situations three examples have been selected. These are:

1. The probability of exceeding certain low temperatures critical for some common agricultural crops during the spring season,

2. The variation in home heating requirements with nocturnal temperature, and

3. The prediction of areas with the greatest probability of nighttime ground fog.

The purpose of these examples is to show how knowledge of the pattern of nocturnal temperature can be used to investigate or predict the areal variation in other, temperature related phenomena; or the suitability or a location for a particular type of land use. Preliminary climatic studies of an area would allow for more rational land use decisions and eliminate the wasted effort and expense involved in climatically non-compatible land uses. The examples presented here are limited to only one climatic element (nighttime temperature), involve very simple methodology, and are merely selected to show the kind of thing that can be done once the relevant climatic data is obtained. Even so, much of what is revealed in an abstract analysis is born out by practical experience on the Saanich Peninsula.

Probability of Occurrence of Critical Spring Temperatures

Every crop plant has a critical minimum temperature below which severe damage will occur or future growth and yield will be markedly reduced. The actual critical temperature varies with each plant species and with different stages of the life cycle of the plant. It is essential that

109

varieties be selected for a particular region which may be planted, go through the successive stages of their life cycle, and be harvested without encountering an unacceptable probability of being damaged by the occurrence of the critical minimum temperature at any stage of their development. Long term temperature records allow the probability of occurrence of any critical temperature to be calculated for any time period. A knowledge of the probability of reaching certain critical temperatures allows the farmer to select those crop plants which provide the best balance between final monetary return and risk of crop loss through cold damage. It also allows him to select planting dates that offer the best chance of avoiding the risk of critical damaging temperatures both in the spring and in the fall before the crop is ready for harvest. The following example looks at the probability of occurrence of certain critical temperatures for a variety of crops during the spring season.

The probability of occurrence of a number of critical temperatures was computed for each week from March 1 to May 30 using long term climatic data recorded at Victoria International Airport. Probabilities were then re-calculated for areas with average clear night temperatures above or below those of the airport. Calculations were made for even degrees Celsius from three degrees above to three degrees below the airport temperature corresponding to the isolines on the temperature distribution map (Figure 2, 3).

Probability of Frost

The probability of recording temperatures below freezing at standard screen height varies greatly over the Saanich Peninsula (Table 2, 3). The airport has a high risk of screen frost until the first week in April. The frost probability remains constant at about .40 during the last three weeks of April and then declines very rapidly to where there is little chance of frost in May. A high risk of frost persists in the frost pocket areas of the Peninsula until early May and some risk continues well into June. The warmer ridges and hilltops, however, have very little chance of screen frost after the beginning of April. Areas located on the + 2° C isanomaly can expect about six weeks more of frost free conditions in an average spring than can areas along the - 2° C isoline.

Although screen frosts are widely discussed and are frequently the only measure of frost occurrence available in published climatic data, it is the temperature much nearer the ground which has a direct effect on newly emergent plants during the spring. The air temperature in the first few inches above the ground surface can be several degrees colder on clear nights than those at screen height. The probability of frost near the surface (ground frost) is, therefore, much greater than the probability of screen frost.

The probability of ground frost throughout the spring season is presented in Table 3.3. The 3° C temperature difference found to occur between the standard screen temperature and height of 2.5 inches (6 cm) on clear nights during the spring of 1975 was applied in this analysis.

The chance of ground frost is much higher than that for screen frost for any given week. The frost risk remains appreciable throughout the month of May in all but the warmer areas of the Saanich Peninsula. The valley bottom areas should plan on the virtual certainty of ground frost until at least the latter half of May. Low growing crops which are susceptible to brief periods of light frost are suitable for only the warmest areas of the Peninsula and even then planting should be delayed so that the young plants are not fully emergent until early May.

The simple probability of frost is not a completely reliable indication of the suitability of a particular area for crop growth. Most mid-latitude crop plants have critical temperatures which are below the freezing point.[18] The extent of damage is also dependent on the degree to which the temperature falls below the critical level, the duration of the

TABLE 2,3

WEEKLY PROBABILITY OF SCREEN FROST
DURING THE SPRING SEASON

Week	Temperature zones deviation from airport temperature ($^{\circ}$C)						
	+3	+2	+1	0	−1	−2	−3
March							
1 - 7	.63	.75	.88	.97	1.0	1.0	1.0
8 - 14	.41	.66	.75	.91	1.0	1.0	1.0
15 - 21	.22	.41	.53	.81	.97	1.0	1.0
22 - 28	.19	.53	.63	.84	.91	1.0	1.0
March							
29 - April 4	.10	.32	.48	.74	.94	.97	1.0
April							
5 - 11	.03	.06	.16	.41	.94	.97	.97
12 - 18	.03	.06	.16	.41	.63	.81	.94
19 - 25	0	.06	.13	.44	.72	.81	.94
April							
26 - May 2	0	.09	.19	.41	.56	.72	.84
May							
3 - 9	0	0	.03	.13	.31	.47	.75
10 - 16	0	0	0	0	.06	.25	.50
17 - 23	0	0	0	0	.06	.25	.34
24 - 30	0	0	0	0	.06	.09	.25

TABLE 3,3

WEEKLY PROBABILITY OF GROUND FROST DURING THE SPRING SEASON

Week	Temperature zones deviation from airport temperature (^oC)						
	+3	+2	+1	0	-1	-2	-3
March							
1 - 7	.97	1.0	1.0	1.0	1.0	1.0	1.0
8 - 14	.91	1.0	1.0	1.0	1.0	1.0	1.0
15 - 21	.81	.97	1.0	1.0	1.0	1.0	1.0
22 - 28	.84	.91	1.0	1.0	1.0	1.0	1.0
March							
29 - April 4	.74	.94	.97	1.0	1.0	1.0	1.0
April							
5 - 11	.41	.94	.97	.97	1.0	1.0	1.0
12 - 18	.41	.63	.81	.94	1.0	1.0	1.0
19 - 25	.44	.72	.81	.94	1.0	1.0	1.0
April							
26 - May 2	.41	.56	.72	.84	.91	.97	1.0
May							
3 - 9	.13	.31	.47	.75	.84	.94	1.0
10 - 16	0	.06	.25	.50	.72	.91	.94
17 - 23	0	.06	.25	.34	.34	.69	.84
24 - 30	0	.06	.09	.25	.34	.50	.81

critical temperature, and the condition of the plant; as well as other
biological, soil and climatic factors. These latter factors are very difficult to consider in a general analysis. Determination of the probability
of achieving temperatures below the freezing point critical for specific
plant species is, however, simply an extension of the procedure for determining the risk of frost.

Probability of Occurrence of Critical Springtime Temperatures for Tree Fruits

Fruit trees are especially susceptible to damage by low temperatures
during the blossom stage in spring when the proper combination of plant
sensitivity and seasonally cold nights is most likely to occur. The average
critical blossom stage temperature in even degrees is -2°C for apples and
-1°C for cherries and plums.[19] Murcier[20] reports that most tree fruits in
full blossom will suffer at least light damage when temperatures fall below
about -2°C.

The risk of achieving the critical temperature for most tree fruits
during the blossom stage in spring is slight over most of the Saanich Peninsula (Tables 4,3 and 5,3). Areas having clear night temperatures equal
to or warmer than the airport offer little risk of damage. The frost pocket
areas, however, can expect at least a 40 per cent chance of damage to
more temperature sensitive fruits (i.e. cherries and plums) and at least a
30 per cent chance of damage to apples until early May.

It is obvious that the warmer upper slope regions are the areas most
suitable for fruit crops in terms of nocturnal temperature. The valley bottoms
have an unacceptably high risk of springtime cold damage to be viable for
commercial production. The lower slopes which have temperatures slightly
below those of the airport are more suitable for apples or other more cold
tolerant crops than for temperature sensitive species. The time of blossoming is also critical. Trees that blossom later will be more suitable than
will those that blossom earlier. Plums, for example, blossom earlier than

114

PLATE 4
Fruit orchard on the Saanich Peninsula.

TABLE 4,3

WEEKLY PROBABILITY OF OCCURRENCE OF THE CRITICAL
TEMPERATURE FOR APPLES DURING THE BLOSSOM STAGE (-2°C)

Week	Temperature zone deviation from airport temperature (°C)						
	+3	+2	+1	0	-1	-2	-3
March 29 – April 4	0	0	.03	.10	.32	.68	.74
April 5 – 11	0	0	.03	.03	.06	.38	.41
12 – 18	0	0	0	.03	.06	.34	.41
19 – 25	0	0	0	0	.06	.28	.44
April 26 – May 2	0	0	0	0	.09	.31	.41
May 3 – 9	0	0	0	0	0	.03	.13
10 – 16	0	0	0	0	0	0	0
17 – 23	0	0	0	0	0	0	0

TABLE 5,3

WEEKLY PROBABILITY OF OCCURRENCE OF THE CRITICAL TEMPERATURE
FOR CHERRIES AND PLUMS DURING THE BLOSSOM STAGE (-1°C)

Week	Temperature zone deviation from airport temperature (°C)						
	+3	+2	+1	0	-1	-2	-3
March 29 – April 4	0	.03	.10	.32	.68	.74	.94
April 5 – 11	0	.03	.03	.06	.38	.41	.94
12 – 18	0	0	.03	.06	.34	.41	.63
19 – 25	0	0	0	.06	.28	.44	.72
April 26 – May 2	0	0	0	.09	.31	.41	.56
May 3 – 9	0	0	0	0	.03	.13	.31
10 – 16	0	0	0	0	0	0	.06
17 – 23	0	0	0	0	0	0	.06

cherries and are, therefore, not as viable on the Saanich Peninsula. In no case would the Saanich Peninsula, the "banana belt" of Canada, be suitable for the growing of bananas.[21]

Probability of Occurrence of Critical Temperatures for Small Fruits and Vegetables

Newly emergent and low growing fruit and vegetable crops will be subjected to the colder air temperatures which occur in the first few inches above the ground. Many vegetable crops are very tolerant to cold temperatures during the early stages of their life cycle. Others are damaged by only of degree or two of frost. Critical early season temperatures for some of the crops widely grown on the Saanich Peninsula are: beans, cabbage, carrots, cauliflower, and turnips, $-6^{\circ}C$; cucumbers, tomatoes, and strawberries, $-1^{\circ}C$.[22] The former group can be considered to be cold tolerant and the latter group to be rather susceptible to cold damage.

Tables 6,3 and 7,3 present the probability of exceeding the critical cold temperatures for the tolerant and susceptible plant groups at a height of 2.5 inches (6 cm). All of the Saanich Peninsula, except the very bottom of the Tod Creek Valley located within the $-3^{\circ}C$ isonomaly, is suitable for hardy vegetable crops that emerge after the beginning of April. The warmer zones clearly offer no problem as far as surpassing the $-6^{\circ}C$ critical temperature. In these areas other factors, such as soil temperature necessary for germination, or hours of sunlight, are more of a limiting factor as to when the crop can be planted. Tender plants, however, are susceptible to cold damage well into the month of June in the low lying frost pocket areas. Even on the warmer slopes the risk of critically cold temperatures remains high until the beginning of May. Late planting of fast maturing varieties or frost prevention methods would be necessary for the reliable production of temperature sensitive, low growing fruit and vegetable crops.

TABLE 6,3

WEEKLY PROBABILITY OF OCCURRENCE OF CRITICAL TEMPERATURE FOR COLD TOLERANT VEGETABLE CROPS (-6° C)

Week	Temperature zone deviation from airport temperature (°C)						
	+3	+2	+1	0	-1	-2	-3
March							
1 - 7	.09	.09	.09	.31	.63	.75	.94
8 - 14	.03	.03	.13	.22	.41	.66	.84
15 - 21	0	0	0	.13	.22	.41	.78
22 - 28	0	0	0	0	.19	.53	.72
March							
29 - April 4	0	0	0	.03	.10	.32	.68
April							
5 - 11	0	0	0	.03	.03	.06	.38
12 - 18	0	0	0	0	.03	.06	.34
19 - 25	0	0	0	0	0	.06	.28
April							
26 - May 2	0	0	0	0	0	.09	.31
May							
3 - 9	0	0	0	0	0	0	.03
10 - 16	0	0	0	0	0	0	0

TABLE 7,3

WEEKLY PROBABILITY OF OCCURRENCE OF CRITICAL TEMPERATURE FOR FROST SUSCEPTIBLE CROP PLANTS (-1° C)

Week	Temperature zone deviation from airport temperature (°C)						
	+3	+2	+1	0	-1	-2	-3
March							
1 - 7	.75	.94	.97	1.0	1.0	1.0	1.0
8 - 14	.66	.84	.91	1.0	1.0	1.0	1.0
15 - 21	.41	.78	.81	.97	1.0	1.0	1.0
22 - 28	.53	.72	.84	.91	1.0	1.0	1.0
March							
29 - April 4	.32	.68	.74	.94	.97	1.0	1.0
April							
5 - 11	.06	.38	.41	.94	.97	.97	1.0
12 - 18	.06	.34	.41	.63	.81	.94	1.0
19 - 25	.06	.28	.44	.72	.81	.94	1.0
April							
26 - May 2	.09	.31	.41	.56	.72	.84	.94
May							
3 - 9	0	.03	.31	.31	.47	.75	.88
10 - 16	0	0	0	.06	.25	.50	.88
17 - 23	0	0	0	.06	.25	.34	.63
24 - 30	0	0	0	.06	.09	.25	.38

Agricultural Land Use in Relation to Nocturnal Temperature

The agricultural land use pattern on the Saanich Peninsula shows a good adjustment to clear night nocturnal temperature patterns. This, in spite of the fact that current land use represents the adjustment to a number of social, economic, pedologic, hydrologic and other climatic factors.

The predominant crops grown on the Saanich Peninsula are those which are more cold tolerant. Apples, for instance, are by far the most important tree fruit and are the only one to show a noticeable expansion over the last fifteen years. 950,000 lbs. of apples were produced in 1973 compared with 792,000 lbs. in 1958.[23]   The production of cherries declined from 50,000 lbs. to 16,000 lbs. during the same period. Often tree fruits are grown only for local consumption.

Leading vegetable crops are cabbage, turnips, carrots, onions and cauliflower, all of which can do quite well in the colder, low lying areas of the peninsula. The production of all but turnips increased from 1958 to 1973. The production of field tomatoes and cucumbers is relatively small and declined during the 1958-1973 period. These crops are, perhaps, better suited for greenhouse production on the Saanich Peninsula.

One of the factors in the changing pattern of agricultural production on the Saanich Peninsula is the impact of urban development. Urban expansion has been concentrated more in the favourable, warmer slope regions of the peninsula and thus reduced the amount of land suitable for temperature sensitive crops. The cold, often damp valley bottoms are not favoured by either urban development or agriculture and the future composition of agricultural output will be dominated more by crops that can grow well in the colder lowlands.

An interesting comparison can be made between the 1973 land use map of the Saanich Peninsula[24] and the map of the nocturnal temperature patterns. Land used to grow fruit, berry and flower crops is concentrated in the warmer regions of the Peninsula. The percentage of contiguous fields growing fruit, berries, flowers, bulbs, holley and christmas trees located in each of the temperature zones is: -2 to -1° C, 0 per cent; -1 to 0° C, 10 per cent; 0 to +1° C, 53 per cent; +1 to +2° C, 33 per cent and warmer than +2° C, 4 per cent. The concentration of production in the 0 to +1° C area allows the advantage of warmer nighttime temperatures without some of the problems of steep slopes, poor soils and lack of water found in some of the warmer areas at higher elevations.

Field vegetables are found over a wide range of temperature zones 20 per cent of the contiguous fields are located in the -2° to -1°C region, 10 per cent in -1 to 0° C; 30 per cent in 0 to +1° C and 40 per cent in +1 to 2° C. The lower critical temperatures for a number of the field vegetables grown on the Saanich Peninsula, plus the greater flexibility of planting dates that allow the avoidance of critical temperatures in the early spring can partially account for their occurrence in even some of the frost pocket areas of the Peninsula.

The dominant land use in the frost pocket areas of the peninsula is pasture. Pasture grasses are little affected by frost and represent a good productive use of the colder areas. The percentage of land use in the areas within the -2° C isanomaly is: improved pasture, 40 per cent; unimproved pasture, 25 per cent; forest, 21 per cent; field vegetables, 10 per cent; and other, 4 per cent.

Thus, land use on the Saanich Peninsula seems well adapted to nighttime minimum temperatures as well as to the other physical and economic factors operating in the region.

Heating Degree Days and Home Heating Fuel Requirements

The amount of heating needed in order to maintain comfortable conditions inside a home is dependent on both building construction and climate. Although a number of climatic elements are involved it has been found that the amount of required heating fuel can be predicted simply as a function of air temperature. The measure of temperature used to estimate fuel requirements is the heating degree day which is the sum of mean daily temperatures below the critical temperature at which home heating becomes necessary.[25] In North America, for example, to maintain the interior temperature of a home at $21.1°$ C $(70°$ F) heating is required when the outside mean daily air temperature drops below $18.3°$ C $(65°$ F). $21.1°$ C is taken as the interior temperature comfortable for North Americans. The critical base temperature is somewhat below the preferable indoor temperature to allow for such things as energy gain from solar radiation, the heat production of indoor processes such as cooking and lighting, and the damping effect that the house has on natural temperature cycles. Heating degree days are calculated by accumulating the number of degrees the mean daily temperature falls below $18.3°$ C for each day throughout the year or heating season. If, for example, the mean daily temperature on a particular day is $10°$ C $(50°$ F) the number of Celsius degree days would be $18.3 - 10.0 = 8.3$. Similarly the number of Fahrenheit degree days would be $65 - 50 = 15$.

Victoria International Airport averages 3167 C degree days (5699 F degree days) for the year as a whole. The monthly averages range from 76 C degree days (136 F) in July to 467 C degree days (840 F) in January.

Maunder[26] examined a limited sample of Victoria homes over the period from December 1965 to April 1967 and found that each C degree day cost the homeowner .245 gallons (1.11 litres) of home heating fuel.

A two degree change in nocturnal minimum temperature will change the mean daily temperature one degree and, thus, add or subtract one heating degree day when the mean daily temperature is below

18.3° C. Considering just the average number of clear nights in each month for which the temperature observations in this study are valid, the change in degree days over any period can be computed. Those homes situated on the $\pm$ 2° C isanomaly line would accumulate $\pm$ 38 heating degree days compared to homes located on the 0° C line. During the spring season this would change fuel requirements by 9.3 gallons which at the spring 1975 retail price of $.375 per gallon would change home heating costs by $2.49. The changes over the whole year would be: 111 C degree days, 27.2 gallons of fuel and $9.92. Changes for those people living on the $\pm$ 1° C isanomaly would be half these amounts.

The above values are not large. This is primarily the result of considering only clear nights which are rather infrequent during the bulk of the heating season. The above illustrates the method, however, which could be applied to nights with all types of weather conditions if the proper temperature data were available. Limited observations show that the valley areas also have somewhat lower temperatures on cloudy nights although the magnitude is small and the pattern quite variable. Another area where more data is needed is in the variation of daytime maximum temperature which has an equal weight in the determination of heating degree days. Temperature transects run in the Victoria area during the summer show that areas under the influence of the sea breeze can be up to 7 or 8° C cooler than more protected inland areas during the afternoon. When large elevation differences are involved valley bottoms are usually warmer than upper slopes. During periods of limited soil moisture, however, the moister valley bottoms are often cooler because of higher evaporation rates. Differences in exposure to solar radiation also makes any generalization about spatial variation in daytime maximum temperature difficult without an accurate data base.

Prediction of Areas of Ground Fog Formation

Fog is defined as a cloud whose base is at the earth's surface and which is sufficiently dense to reduce visibility to less than one kilometre. The practical significance of fog is chiefly its interruption of transportation.

In theory, condensation will occur when the temperature of the lower layers of the atmosphere falls below the dew point temperature. The actual critical temperature, however, may be above or below the dew point depending on the abundance and efficiency of condensation nuclei which are present. The U. S. Civil Aeronautics Administration[27] says that aviators should consider the development of fog or low cloud a possibility whenever the air temperature is within 4° F (2.22° C) of the dew point. Petterssen[28] in a discussion of fog forecasting, however, emphasizes that additional cooling below the dew point is often necessary in order to produce a sufficient liquid water density to reduce visibility below one kilometre. He suggests that a moderate fog would contain at least 0.5 grams of liquid water per cubic metre of air. The additional cooling below the dew point required to achieve this liquid water density varies with temperature from about 0.4° C at 20° C to 2.0° C at -5.0° C.

On the ten clear nights during which observations were conducted in the spring, the air temperature at the airport averaged 1.3° C above the dew point temperature. The difference on individual nights ranged from 0 to 2.8° C. Eliminating the night on which the air temperature equalled the dew point a great deal of variation exists around the Saanich Peninsula in the frequency with which measured air temperature was below the airport dew point. A total of 50 out of 221 individual observations taken at a variety of locations over the nine nights had temperatures below the dew point recorded at the airport. 36 per cent of the below dew point observations occurred in the Tod Creek Valley and 34 per cent occurred in the Hagan Creek Valley. Individual stations in the Tod Creek Valley had temperatures below the dew point on between 60 and 100 per cent of the observation nights. Individual stations on the margins of the Hagan Creek Valley had frequencies varying between 0 and 78 per cent. Most of the rest of the below dew point observations were located in the Stelley's Cross Road-Wallace Drive Lowland and the McHugh Valley.

An average of 1.4° C further cooling below the dew point would be needed to produce a fog density of 0.5 gm/m^3 and reduce visibility below one kilometre. The value on individual nights ranged between 0.9° C and 2.1° C. Only 14 of the observations had temperatures below this critical value. Four of these were in the Hagan Creek Valley The Tod Creek Valley and the Stelley's Cross Road-Wallace Drive Lowland had five each.

In summary, it is quite apparent that the colder an area the greater the probability for ground fog. In this analysis advection fog from the sea is not considered so it is the valley bottom areas which are most likely to suffer the disrupting effects of ground fog. The air temperature was sufficiently below the dew point recorded at the airport for fog formation to be likely at one or more observation point on about one-half of the springtime observation nights. All these stations were located in the valley bottoms. In practise, ground fog was noticed on about one clear night in four on which transect runs were undertaken throughout the year. Although fog could have been present and gone unnoticed if it was not immediately adjacent to the road, all fogs were in the lowest valley bottoms. It is these areas which should be avoided by any land use which is sensitive to the problems caused by low visibility.

SUMMARY

The Saanich Peninsula with its variety of surface form, complex mixture of land and water surfaces and diversity of land use provides an ideal laboratory for the study of the interrelationship between topography and climate. In the case of temperature on clear, radiation nights it is land surface form which is the most important factor in determining the local temperature pattern. The valley bottoms are almost textbook examples of the classic frost pocket. Warmer nighttime temperatures are found on the upper slopes which have free air drainage. The sea and inland lakes have a smaller but noticeable effect in moderating temperatures.

Three examples were presented to illustrate how even very simple methodology may be applied to assess the practical significance of local climatic variation. A more complete study would have to involve many more climatic elements. Despite the limitation of this study to one element, however, much of what is suggested by theoretical analysis is found to actually occur in the real world. Agricultural land use, for instance, has adjusted itself to nocturnal temperatures as well as to the whole host of other physical constraints found on the Peninsula. The more temperature sensitive agricultural activities are located in the warmer areas and the frost pockets are reserved for pasture, hardy field vegetables or forest.

In most cases the adjustment of land use to climate occurs only after a long period of trial and error. It is hoped that the present study has demonstrated one method whereby short term observations can be tied to a climatic station with a long term record. This information can then be used to map the local variations in climate and provide a basis on which the most successful types of land use can be selected.

REFERENCES

1. GRIGORYEV, A. A. and BUDYKO, M. I.
   "Classification of the Climates of the U. S. S. R." Soviet
   Geography, Review and Translation   1, No. 4/5
   (1960), pp. 3 - 24.

2. MONKHOUSE, F. J. A Dictionary of Geography. London:
   Edward Arnold, 1965, p. 311.

3. For a more complete discussion of the physical, biological and
   land use conditions on the Saanich Peninsula see:
   STANLEY - JONES, C. V. and BENSON, A. W. (eds.)
   An Inventory of Land Resources and Resource Potentials
   in the Capital Regional District. Report prepared for the
   Capital Regional District, 1973.

4. HAWKE, E. L. "Thermal Characteristics of a Hertfordshire Frost-
   hollow," Quarterly Journal of the Royal Meteorological
   Society, 70, (1944), pp. 23-40; LAWRENCE, E. N.
   "Temperatures and Topography on Radiation Nights,"
   Meteorlogical Magazine, 87, No. 1029 (1958), pp.
   71-75. "Minimum Temperature and Topography in a
   Herefordshire Valley, "Meteorlogical Magazine, 85,
   No. 1005 (1956) p. 79 - 83; GEIGER, R. The Climate
   Near the Ground, Cambridge, Massachusetts: Harvard
   University Press, 1966; SMITH, K. "A Note on Minimum
   Screen Temperatures in the Houghall Frost Hollow."
   Meteorological Magazine, 96, No. 1143 (1967), pp. 300-
   302; LONGLEY, R. W. and LOUIS - BYNE, M. Frost
   Hollows in West Central Alberta. Canada Department
   of Transport, Toronto, Circular No. 4532, 1967; HARRISON,
   A. A. "Variations in Night Minimum Temperatures Pec-
   uliar to a Valley in Mid-Kent, "Meteorological Magazine,
   96, No. 1142 (1967), pp. 257 - 265. "Discussion of the
   Temperature of Inland Kent with Particular Reference to
   Night Minima in the Lowlands," Meteorological Magazine,
   100, No. 1185 (1971), pp. 97 - 111; HOCEVAR, A. and
   MARTSOLF, J. D. "Temperature Distribution under Radiation Frost Conditions in a Central Pennsylvania Valley,"
   Agricultural Meteorology, 8, No. 4/5 (1971), pp. 371-
   383; MACHATTIE, L. B. "Kananaskis Valley Temperature
   in Summer," Journal of Applied Meteorology, 9, No. 4
   (1970), pp. 574 - 582.

5. HERLINVEAUX, R. H. "Oceanography of Saanich Inlet in Van-
      couver Island, B. C.," Journal of the Fisheries Research
      Board of Canada, 19, No. 1 (1962).

6. HARRISON, 1971, op. cit.

7. LAWRENCE, 1958, op. cit.

8. HOCEVAR and MARTSOLF, op. cit.

9. HOCEVAR and MARTSOLF, op.cit.; HARRISON, 1967, op. cit.

10. For a more complete discussion of the urban heat island see SUNDBORG,
      A. "Local Climatological Studies of the Temperature
      Conditions in an Urban Area," Tellus, 2, No. 3 (1950),
      pp. 222 - 232; DUCKWORTH, F. S. and SANDBERG,
      J. S. "The Effect of Cities upon Horizontal and Vertical
      Temperature Gradients," Bulletin of the American Met-
      eorological Society, 35, No. 5 (1954), pp. 198 - 207;
      LANDSBERG, H. E. "The Climate of Towns," in
      THOMAS, W. L. (ed.) Man's Role in Changing the Face
      of the Earth. Chicago: University of Chicago Press,
      1956. PARRY, M. "Local Temperature Variations in the
      Reading Area," Quarterly Journal of the Royal Meteoro-
      logical Society, 82, No. 351 (1956), pp. 45 - 57;
      PETERSON, J. T. "Climate of the City, in DETWYLER,
      T. R. (ed.) Man's Impact on Environment. New York:
      McCraw - Hill, 1971.

11. DUCKWORTH and SANDBERG, op. cit.; FUKUI, E. "Increasing
      Temperature Due to the Expansion of Urban Areas in Japan,"
      Journal of the Meteorological Society of Japan, 75, (1957),
      pp. 336 - 341; CHANDLER, T. J. "City Growth and
      Urban Climates," Weather, 19, No. 6 (1964), pp. 170-
      171; SUMMERS, P. W. An Urban Ventilation Model
      Applied to Montreal. Ph.D. Dissertation, McGill
      University, Montreal, 1964; OKE, T. R. "City Size and
      the Urban Heat Island," Atmospheric Environment, 7, No. 8,
      (1973), pp. 769 - 779.

12. NORWINE, J. R. "Heat Island Properties of an Enclosed Multi-
      level Suburban Shopping Center," Bulletin of the
      American Meteorological Society, 54, No. 7 (1973),
      pp. 637 - 641.

13. OKE, T. R. and EAST, C. "The Urban Boundary Layer in Montreal." Boundary-Layer Meteorology , 1, No. 4 (1971), pp. 411 - 437.

14. TULLER, S. E. "Nighttime Urban Climatic Variations and Their Implications for Human Thermal Comfort in Different Circum-Pacific Climatic Regimes," Proceedings I. G. U. Regional Conference, Palmerston North, New Zealand, (forthcoming).

15. For instance see KINGHAM, H. H. "Surface Temperature Patterns in Christchurch at Night," New Zealand Geographer, 25, No. 1 (1969), pp. 16 - 22; FONDA, R. W. et. al. "Heat Islands and Frost Pockets in Bellingham, Washington," Bulletin of the American Meteorological Society, 52, No. 7 (1971), pp. 552 - 555.

16. FONDA et. al., op. cit.

17. HAWKE, op. cit.

18. CHANG, J. H. Climate and Agriculture. Chicago: Aldine, 1968.

19. SHAW, R. H., THOM, H. C. S. and BARGER, G. C. "The Occurrence of Freezing Temperatures in the Spring and Fall," in IOWA STATE COLLEGE OF AGRICULTURE AND MECHANICAL ARTS, The Climate of Iowa. AMES: IOWA STATE COLLEGE, 1952.

20. MURCIER, R. G. Protection Against Frost in Southern Ontario. Vineland Station, Ontario; Horticultural Experiment Station,

21. Personal communication, Ian F. Owens, Department of Geography, University of Canterbury, Christchurch, New Zealand, May 7, 1975.

22. CHANG, op. cit.; SHAW et. al., op. cit. and MURCIER, op. cit.

23. Statistics on agricultural production and part of the discussion of trends was taken from BRITISH COLUMBIA DEPARTMENT OF AGRICULTURE, Agriculture Outlook - A Short and Long Term Outlook for Vancouver Island. Victoria: B. C. Dept. of Agriculture, 1972.

24. STANLEY - JONES and BENSON, op. cit.

25. For a further discussion of heating degree days see: CONRAD, V.
    and POLLAK, L. W. Methods in Climatology. Cambridge,
    Massachusetts: Harvard University Press, 1950;
    AMERICAN SOCIETY OF HEATING, REFRIGERATION AND
    AIR CONDITIONING ENGINEERS, ASHRAE Guide and
    Data Book: Applications. NEW YORK: ASHRAE, 1964,
    pp. 215 - 226; BOYD, D. W. Climatic Information For
    Building Design in Canada. Ottawa: National Research
    Council, NRC No. 11153, 1970; WILLIAMS, G. D.
    and MACKAY, K. H. Tables of Daily Degree Days Above
    or Below Any Base Temperature. Canada Dept. of Agri-
    culture, Publication No. 1409, Ottawa: Queen's Printer,
    1970.

26. MAUNDER, W. J. The Value of the Weather. London: Methuen,
    1970.

27. CIVIL AERONAUTICS ADMINISTRATION. Pilot's Weather Hand-
    book. CAA Technical Manual No. 104, Washington,
    D. C.: U. S. Government Printing Office, 1955.

28. PETTERSSEN, S. Weather Analysis and Forecasting. New York:
    McGraw - Hill, 1940.

CHAPTER 4

COASTAL EROSION: A NATURAL HAZARD
OF THE SAANICH PENINSULA, VANCOUVER ISLAND

Harold D. Foster

University of Victoria

AN INTRODUCTION TO RISK [1]

Despite the ubiquitous desire for safety, disasters are common-place.[2] This paradox is largely the result of an insufficient appreciation of the nature of risk. Risk, in this context, is defined as the chance that an adverse event, such as a flood, earthquake or fire, may occur during some specified time period. It is measured as a probability and differs greatly from location to location.[3] This is because the operation of meteorological and geomorphological processes varies spatially, both in intensity and frequency and the earth's surface is, as a result, an intricate risk mosaic. Unless such areal differences in risk are fully appreciated, competent decision-making is greatly hindered. Yet this is commonly the case. All too frequently, ignorance about the risks involved in developing an area results in construction on vulnerable sites, creating a hazard which may eventually lead to disaster.[4] There is considerable evidence that this is the case with coastal erosion on the Saanich Peninsula, Vancouver Island.

A further complicating factor is frequent misunderstanding over responsibilities associated with risk taking. Where accountability is fragmented, or denied by several levels of government, or where those directly at risk anticipate financial compensation should disaster strike, the legality and frequency of higher risk taking increases.[5]

The hazards associated with varying rates of coastal erosion on the Saanich Peninsula, Vancouver Island are the focus of this study. These are established by a detailed description of the geomorphological processes invoved and the lives and properties at risk. The probability of disasters resulting from this cause are then discussed and suggestions made which, if implemented, could greatly reduce possible losses of life and property.

HAZARD PROPENSITY IN
COASTAL ZONES

The physical and social environments are interrelated through a series of sub-systems. A change in the physical milieu, such as an acceleration in the rate of coastal erosion, may commonly precipitate a social response, for example the construction of sea walls. Similarly social change, illustrated by the building of groynes, may disturb the dynamic equilibrium of shoreline processes, perhaps causing a redistribution of coastal erosion and deposition.

Social and physical systems are, therefore, interlinked, each responding to change in the other through a series of negative and positive feedback loops.[6] Where the rate of change intensifies in either system, (unless it has been deliberately stimulated as an aid to disaster mitigation), risks to life and property tend to increase. This is because it is in these areas of rapid change that the interrelationship is in the greatest disharmony, neither system having had sufficient time to reestablish equilibrium. Rapid change, therefore, generally implies high risk.

Physical Variables

Change is frequently occurring along the coast because it is here that an essentially unidirectional flow of energy from the sea interacts with the shoreline. This continual flow of wave, current, tidal and wind energy against the coast is mainly consumed in eroding the cliffs and redistributing the resulting beach sediments. The net tendency is to planate the land-sea interface and to remove headlands and fill bays, so producing a linear shoreline which retreats landwards. However, as Grant et. al. have pointed out:

> . . . . since perfect linearity is a theoretical
> abstraction, and since shoreline regression can
> never keep pace with changes in base level, the
> coast is continually adjusting to maintain a
> steady-state condition of metastable equilibrium.
> In maintaining such a condition the system is
> infinitely responsive to the multitude of short-
> term and local changes in each variable. The
> sediment load, in transit from source to sink and
> manifest as depositional forms, acts as a very
> sensitive buffer to absorb the energy and reaction
> imbalances created by changing regimes. Hence
> shoreline features, such as spits, barriers, deltas
> and forelands are characteristically changeable and
> ephemeral. Yet here, because of the intrinsic
> aesthetic appeal and conducive physical character-
> istics of these features, man is to be found in
> greatest numbers, modifying and exploiting these
> sensitive and delicate substrates to his own use
> and commonly, misuse.[7]

Along the cliffs energy is expended in erosion; in the surf zone most of it is consumed by turbulence and friction and attrition of beach materials. The refraction of waves at the coast, which causes them to focus most of their energy onto headlands, establishes an energy gradient which decreases laterally towards nearby bays. Also, a small residual force vector is left after wave destruction on the beach, this is directed down the energy gradient, creating a longshore current capable of transporting sediment. Additional material is carried by littoral drifting, the zig-zag movement of beach material induced by oblique wave incidence (Figure 1,4).

Such currents lose energy by friction with sea bed, impact between particles and eddy diffusion with offshore water. In this manner, transport capacity declines and deposition increases, causing the formation of sandspits, deltas and forelands. As Grant et. al. point out, on-shore currents ultimately meet at bayheads where they are deflected seawards as rip currents or sheet underflow.[8] Such currents carry excess sediments offshore, where deposition occurs.

The rates at which coasts actually erode vary markedly, depending upon two distinct groups of variables, termed passive and active factors. Passive factors, such as geology, topography, and the orientation of the coast, and to a lesser extent its vegetation cover do not alter rapidly through time, but are important, however, since they provide the stage upon which the kaleidoscopic marine drama is enacted. In contrast, some significant variables, such as sea state, wave spectrum, wind direction and force, storm path and the height of the tide, alter rapidly and are, therefore, termed active variables. They are the Thespians of the coastal duologue.

Passive factors

(I) Geology

Rock types of differing ages, hardness and solubility are exposed to marine action on the Saanich Peninsula (Figure 2,4). Such bedrock is generally resistant to erosion and, where not capped by weaker Pleistocene sediments, cliff retreat, as a result, is very slow.

The oldest rocks in the area are members of the Vancouver group,[9] variously identified as dating from the Devonian to Jurassic.[10] The Vancouver group is over 10,000 feet in thickness and is furthur subdivided on the basis of its distribution, lithology and structure into

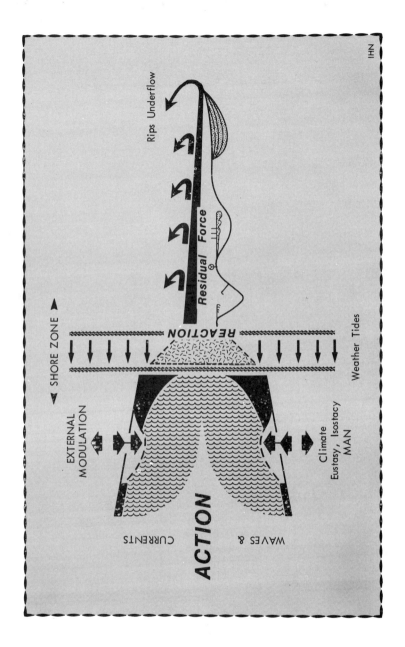

FIGURE 1,4   An Anthro-hydro-geodynamic coastal model.

135

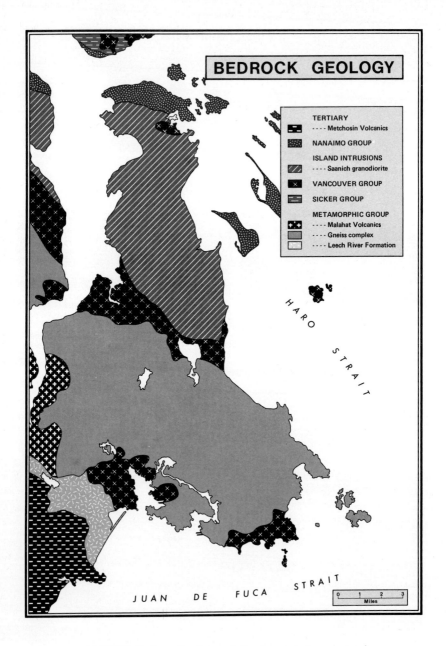

BEDROCK GEOLOGY

TERTIARY
- - - - Metchosin Volcanics

NANAIMO GROUP

ISLAND INTRUSIONS
- - - - Saanich granodiorite

VANCOUVER GROUP

SICKER GROUP

METAMORPHIC GROUP
- - - - Malahat Volcanics
- - - - Gneiss complex
- - - - Leech River Formation

HARO STRAIT

JUAN DE FUCA STRAIT

Miles
0 1 2 3

FIGURE 2,4   Geology of the Saanich Peninsula.

the Vancouver volcanics and the Sutton Formation. The first of these, the Vancouver volcanics predominates. It consists chiefly of andesites with intrusive porphyrites which have been metamorphosed and in some areas (near intrusive granitic rocks) partially recrystallized. Such rocks outcrop discontinuously along the Victorian waterfront from Gonzales Bay to the north of the Oak Bay marina. They are also exposed to marine erosion around portions of Esquimalt Harbour, Cordova Bay, Shoal Harbour in eastern North Saanich, Tod Inlet and Partridge Hill in western Central Saanich. The Vancouver volcanics are resistant to coastal erosion but not immune to it. This distinction is clearly illus-trated by the presence of at least two, perhaps several, preglacial marine abrasion surfaces which have been cut, across the Vancouver volcanics, at heights ranging from present sea level to twenty feet above it. These can clearly be seen along the southern Oak Bay water-front (Figure 3,4).

Intercalated with these volcanics are numerous lentils of crystalline limestones which have also been metamorphised and greatly altered. Known as the Sutton formation such limestone rarely outcrops along the Coast. It can be seen, however, in Tod Inlet.[11] Because of their solubility such limestones are slightly more susceptible to erosion, but rates of retreat are slow and do not constitute a threat to adjacent developments.

The Vancouver group, as a whole, was greatly deformed during the Coast Range Orogeny, (probably during the Jurassic)[12] into large folds with steep limbs, which strike nearly N60°W. These structures are further complicated by smaller folds and contortions, and are also greatly fractured and faulted.

During and after this deformation, plutonic rocks invaded the Vancouver group and now occupy the greater part of the Saanich Peninsula and Victoria area. Clapp divided these rocks into three main

FIGURE 3,4
Pre-glacial marine abrasion surfaces, Oak Bay.

types, the Wark gabbro-diorite-gneiss, Colquitz quartz-diorite-gneiss and Saanich granodiorite. In contrast, Fyles describes the Saanich granodiorite as Island Intrusions of Jurassic or Cretaceous age, while the two remaining plutonic rock types are placed in a Metamorphic Complex of indeterminate age.[13]

The Wark gabbro-diorite gneiss outcrops in Victoria Harbour, Portage Inlet, the Gorge Waterway and Beacon Hill Park. It is also exposed along Ten Mile Point and Gordon Head. In all such areas it shows resistance to present erosion but, nevertheless, has been planated in the past since it now displays a sequence of pre-glacial abrasion surfaces. The lowest of these is being exhumed for example, from beneath Pleistocene surficial sediments at Horseshoe Bay, in Beacon Hill Park (Figure 4,4).

The Colquitz gneiss is a foliated quartz diorite and although intruded separately, forms virtually a single batholith with the Wark gneiss. It is exposed along the coast in Southern Cadboro and Cordova Bays. It also shows a marked resistance to marine erosion, as does the Saanich granodiorite. This latter plutonic rock forms a relatively large batholith and several smaller stocks underlying much of the Saanich Peninsula, where it outcrops on both west and east coasts.[14] Here it shows evidence of pre-glacial marine planation.

Resting unconformably upon these igneous and metamorphic rocks are a series of some 6,000 feet of conglomerates, shales and sandstones known as the Nanaimo Group. These are probably of Upper Cretaceous age, being both terrestrial and marine in origin and are restricted, in this area, to North Saanich and adjacent Islands. These lithologies differ in their resistance to marine erosion and are perhaps slightly more susceptible to landward retreat than the igneous and metamorphic rocks to the south. Within the Nanaimo Group, differing

FIGURE 4.4
Vashon till overlying marine abrasion surfaces, Victoria.

adjacent beds display definite variations in both hardness and their ability to resist marine encroachment, resulting in numerous minor coastal irregularities (Figure 5,4). However, it should be stressed that none of the solid bedrock outcropping along the coast of the Saanich Peninsula is undergoing erosion at a rate that need cause concern in any but the most unusual circumstances.

One further geological factor is of significance to a discussion of coastal erosion in the Saanich Peninsula and Victoria areas. Ten major faults have been indentified, five of which trend northeast – southwest with an equal number orientated northwest – southeast.[15] (Figure 6,4). Bays occur where many of these reach the coast. One, for example, is exposed to marine action at Cole Bay in the west and Roberts Bay in the east, and is perhaps responsible for the formation of both. Similarly, much of the unusual shape of Tod Inlet may be due to the influence of two faults which reach Saanich Inlet at this location. It appears likely that glacial and marine erosion during the Tertiary and Pleistocene, have combined to extend embayments along fault–shattered, more easily eroded bedrock. This process slowly continues.

(2) Topography

The relief of the Saanich Peninsula displays relatively little obvious evidence of bedrock control. Although Mount Newton and Bear Hill (the former rising to over 1,000 feet), are composed of Saanich granodiorite this rock type also underlies the eastern lowlands, (from northern Cordova Bay to Shoal Harbour) and the low lying area between Patricia and Bazan Bays, the site of Victoria International Airport. Similarly the Wark Gabbro diorite gneiss, which, according to Clapp, outcrops at over 1,400 feet on Mount Wark in the Highlands District, is also found at heights below 50 feet in Oak Bay and Beacon Hill Park. The Vancouver volcanics shows similar marked outcrop

FIGURE 5,4
Selective erosion of the Nanaimo Group.

altitudinal variations. The area's relief then cannot be explained simply in terms of the differential erosion of hard and soft rocks.

The Saanich Peninsula forms part of the Nanaimo Lowlands which extend for 175 miles along Vancouver Island's east coast from Sayward on Johnstone Strait to Jordan River, west of Victoria.[16] In the author's opinion, the fundamental difference in relief between the highlands of the west of the Saanich Peninsula and the lowlands of the east is the result of Tertiary and Pleistocene marine erosion. This erosion probably created a number of planation surfaces, a series of step-like levels cut across rocks of varying ages and hardnesses, which have been recognized in numerous other areas at heights of up to at least 700 feet.[17] They are present in the east and south of the Peninsula but not in the west.

The absence of marine Tertiary and Pleistocene erosion surfaces, in the west of the Saanich Peninsula, can be best explained by postulating that Saanich Inlet did not exist as an arm of the sea until after extensive Pleistocene glaciation. Until this time, this region may have been occupied by one of perhaps two rivers carrying drainage from the Highland District and other adjacent mountainous areas. Under these circumstances the present Saanich Peninsula would have been contiguous with Mount Jeffrey and other highland regions, west of what are now Finlayson Arm, Squally Reach and Saanich Inlet. It is suggested that in the late Tertiary, the Saanich Peninsula was a headland extending eastwards into Haro Strait. As a result, higher sea levels during the Tertiary and early Pleistocene bevelled the east and south of what is now the Saanich Peninsula but not the west. Such a distinction would account for the present relief contrast. An alternative explanation is that higher Tertiary and early Pleistocene sea levels were also represented in the west, but their abrasion surfaces were subsequently removed during the glacial widening and deepening of Saanich Inlet.

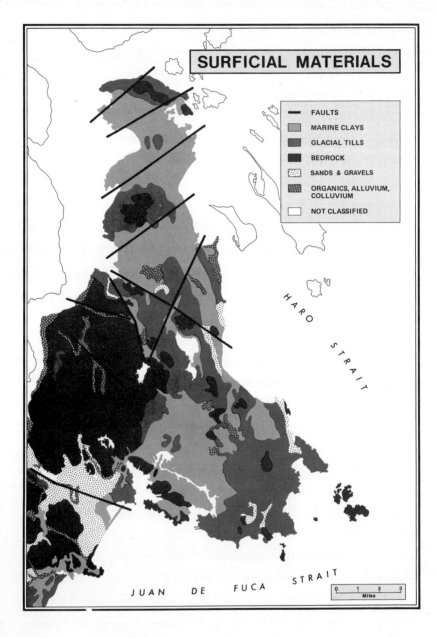

FIGURE 6,4   Quaternary sediments and faults.

144

All of southwestern British Columbia, with the exception of some of the highest mountains, were extensively glaciated.[18] The Saanich Peninsula and Victoria areas were no exceptions.[19] As a result, the Tertiary topography was greatly modified and partially destroyed. Ice probably flowed southwards down the Strait of Georgia from centres of dispersal in the Vancouver Island Ranges and the Interior of British Columbia, at least four times,[20] although local depositional evidence for only the last two occasions, the Dashwood and Vashon glaciations, is convincing.[21]

During both of these glaciations, the ice over the Strait of Georgia was some 5,000 feet thick, reaching approximately 2,500 feet in depth over Southern Vancouver Island.[22] Since the direction of ice flow was basically southwards, what is now the Saanich Peninsula stood as a major obstruction in the paths of these ice tongues. Ice, therefore, flowed both over and around it, probably at higher than average velocities. This ice appears to have been particularly erosive in the west where it was most constricted. Here it deepened and widened what are now Saanich Inlet and Finlayson Arm, escaping southward, between Skirt Mountain and Mount Wells and to the north of Triangular Hill where it created a further depression, now the site of Langford and Glen Lakes.

In the east, the low lying marine abrasion surfaces were not as constrictive and erosion was far less severe, although the bedrock was glacially striated and polished. Numerous roches moutonnées, for example, Mount Douglas, indicate some erosion in the east of the Peninsula.

The evolution of the topography of the Saanich Peninsula has been described in detail since it influences present day marine erosion in a variety of significant ways. The glacial erosion and lowering of Saanich Inlet, Esquimalt Harbour, Victoria Harbour and Portage Inlet has allowed, for example, the sea to penetrate deeply inland and, therefore, exposed a far greater length of coastline to potential marine erosion. Since cliffs are a cross section of the land, their height reflect the relief into which they are cut. The topography also controls the height of the cliffs and, therefore, the volume of material supplied to the sea by landward retreat. Low lying topography, like that resulting from Tertiary erosion around Greater Victoria or parts of eastern Saanich, is not associated with high, steep cliffs. In contrast, glacially steepened valleys, for example Saanich Inlet, are noted for their rugged topography, illustrated by cliffs on the western flank of Jocelyn Hill in Finlayson Arm.

Topography is also very important in controlling the length of fetch of the waves attacking the coasts. Fetch is the distance of open water over which the waves may build up without interference. The greater the fetch, other factors being equal, the higher will be the waves created and, therefore, the faster the rate of coastal erosion.[23]

The lengths of fetch to which the cliffs of the Saanich Peninsula are exposed vary greatly. This variation is illustrated in Figure 14,4. One of the largest calculated length of fetch occurs at Cowichan Head where waves have 51.2 miles of open water over which to build up, should the wind be blowing from the southeast. In contrast, one of the most sheltered locations is Tsehum Harbour which is exposed to a maximum length of fetch of only 1.8 miles from the east. With few exceptions, the south and east coasts of the Saanich Peninsula are more exposed to long fetches than is the western coastline. The potential for their marine erosion is, in consequence, far greater.

(3) Surficial Sediments

The topography of the Saanich Peninsula and the Greater Victoria area, predominantly rugged in the west and low lying in the south and east, significantly influenced the depth, location and nature

146

of Pleistocene deposition. These sedimentary characteristics in turn, now exert a great, perhaps overriding influence on the rates of present coastal erosion.

Southern Vancouver Island was glaciated several times during the Pleistocene era, by ice from the British Columbian mainland and the Island's mountainous interior. Each advance and retreat was associated with a typical sequence of deposits, now exposed to marine erosion. As the ice front advanced southwards, outwash sands and gravels preceded it, being predominantly deposited in low lying areas such as central and eastern Saanich. These sediments were then over-ridden as ice, itself invaded the region. Such ice sheets were most erosive in areas where their flow was impeded by high relief, as in the Highland District. Till deposition occurred where movement was less hampered, as in flat, low-lying areas, for example Beacon Hill Park. Later, as the ice retreated, glacio-fluvial gravels and sands were again deposited, exemplified by the sediments of the Colwood 'delta'. Such outwash was once more deposited peripherally to the upland spine of the Saanich Peninsula, for example to the east, where it formed con-spicuous eskers. These are now being eroded by the sea on Sydney Island, James Island and at Cowichan Head.

Typically, the weight of overriding ice isostatically depressed the Peninsula several hundred feet and, therefore, upon its retreat, the sea invaded the area. Varved marine clays, marking such transgressions, occur in the lower lying parts of the Peninsula. Isostatic rebound, typically rapid, quickly followed, causing a rise in land relative to the sea. As a result terrestrial deposits then accumulated on top of recently exposed marine sediments.

This sequence of events and their associated deposits has occurred twice during the past 60,000 years. The details of the chronology in-volved are illustrated in figures 7,4 and 8,4, which also provide further

147

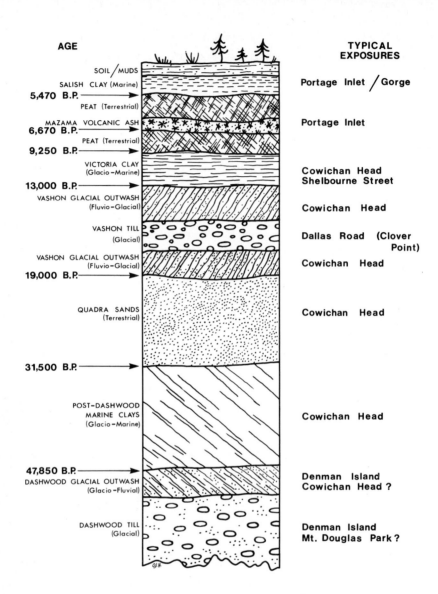

AGE

SOIL/MUDS
SALISH CLAY (Marine)
5,470 B.P.
PEAT (Terrestrial)
MAZAMA VOLCANIC ASH
6,670 B.P.
PEAT (Terrestrial)
9,250 B.P.
VICTORIA CLAY
(Glacio-Marine)
13,000 B.P.
VASHON GLACIAL OUTWASH
(Fluvio-Glacial)
VASHON TILL
(Glacial)
VASHON GLACIAL OUTWASH
(Fluvio-Glacial)
19,000 B.P.
QUADRA SANDS
(Terrestrial)
31,500 B.P.
POST-DASHWOOD
MARINE CLAYS
(Glacio-Marine)
47,850 B.P.
DASHWOOD GLACIAL OUTWASH
(Glacio-Fluvial)
DASHWOOD TILL
(Glacial)

TYPICAL
EXPOSURES

Portage Inlet / Gorge

Portage Inlet

Cowichan Head
Shelbourne Street

Cowichan Head

Dallas Road (Clover
Point)

Cowichan Head

Cowichan Head

Cowichan Head

Denman Island
Cowichan Head ?

Denman Island
Mt. Douglas Park ?

FIGURE 7,4   Quaternary chronology.

148

FIGURE 8,4
Pleistocene and recent sediments,
Cowichan Head.

149

information concerning the physical characteristics of typical sediment. Earlier glaciations also probably took place but are poorly represented in coastal locations.

In summary, Pleistocene topography ensured large scale deposition in low lying areas, such as the eastern Saanich, Victoria and the Colwood areas, while precluding it in highland locations such as Mount Newton and Mount Wark. Although these hills were glacially striated and etched, they were the sites of relatively little Pleistocene deposition. It is, therefore, not surprising that marine erosion is now far more extensive in the south and east, where unconsolidated sediments are exposed to wave action, than in the west, where resistant bedrock predominates along the coast.

(4) Eustatic change

The rates of coastal erosion are clearly influenced by the relative positions of the land and the sea. This relationship is not constant but varies greatly through time. Eustatic change has been studied extensively in southwestern British Columbia by Mathews, Fyles and Nasmith[24] who have derived a series of inferred relative sea level shifts, based on the radiocarbon dating of marine fossils and terrestrial and freshwater organic materials. Evidence of former shoreline locations has also been used to establish the extent of higher sea levels. Their interpretation of sea level changes in the Victoria Area is reproduced as figure 9,4. This diagram is based, to a large degree, on radiocarbon dates obtained by this author from marine and terrestrial deposits exposed by excavation on the northern shore of Portage Inlet.[25]

These dates, and the sediments from which they were taken, indicate that after the retreat of the Vashon ice sheet, from the Saanich Peninsula, circa 13,000 years B. P., relative sea level quickly

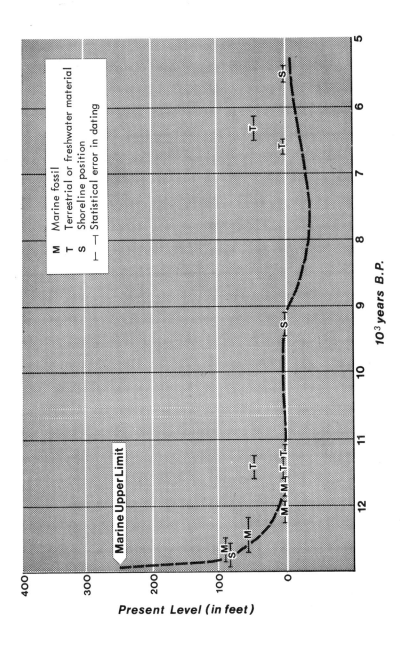

FIGURE 9,4   Sealevel changes in the Victoria area.

151

reached 250 feet higher than it is today. This transgression occurred because of the isostatic depression of the land caused by the former ice cover. During this time Mount Newton, Mount Douglas and Mount Tolmie stood as additional "Gulf Islands", the coast of a reduced Vancouver Island being formed by the highlands of the west of the Saanich Peninsula.

Isostatic rebound occurred quickly, perhaps within 1,000 years, returning the relative relationship of land and sea more or less to its present position. This sea level then remained fairly constant for some 3,000 years. Relative sea level then fell by as much as 35 feet below that of the present. This eustatic depression probably resulted from a glacial readvance elsewhere, which did not reach this area.[26] With the final major retreat of Pleistocene ice from the continents, sea level was reestablished at its present level circa 5,500 years B.P. Since this time it has fluctuated slightly, tidal gauges indicating a current rise of over 0.2 feet per century in the Victoria area.[27]

Sea level then has fluctuated markedly since the last retreat of the ice from the area, being at various times 250 feet above and 35 feet below its present relative position. As a result, the present cliffs probably began their development 12,000 to 9,000 years B.P., when sea level occupied more or less the same relative position it does today. They were then abandoned as sea level fell, but have been subjected to further marine erosion since circa 5,500 years B.P. As a result of such eustatic changes, higher, abandoned cliffs, such as those on the southern flank of Beacon Hill, Victoria (Figure 10,4) are now commonly exposed. Similarly, abandoned but submerged cliffs, dating from the lower sea levels of 9,250 to 5,500 years B.P., can be recognized on marine charts.

FIGURE 10,4
Mass movement below abandoned cliff line, Beacon Hill Park, Victoria.

(5) Vegetation

Typically, extensive vegetation retards coastal erosion. In the Saanich Peninsula, the predominant natural vegetation is a forest-grassland association. The characteristic forest cover is Garry Oak (Quercus garryana), Douglas Fir (Pseudotsuga menziesii) and Arbutus (Arbutus menziesii).[28] This vegetation cover results in leaf canopy interception of rainfall and high transpiration losses, two processes which permit a smaller percentage of the precipitation to enter the ground water system. As a result, less mass movement is induced in coastal surficial sediments. A further influence of vegetation is the binding effects of root systems along cliff edges. In some localities, particularly where the cliffs are relatively low, this may retard the rate at which sediment falls into the sea. However, root systems do not prevent erosion and many of the area's beaches are littered with trees, which have fallen into the sea as a result of coastal erosion.[29]

Active factors

(1) Tidal Range

The Strait of Georgia, in common with much of the Pacific Coast of North America, experiences a mixture of diurnal and semi-diurnal tides.[30] There is a diurnal inequality which affects both tidal times and heights. Its effects principally influence the height and times of succeeding low tides. The Straits of Georgia, therefore, experiences both an approximate two week cycle, as well as a seasonal cycle in tidal ranges. In the Saanich Peninsula, as in other parts of the Straits of Georgia, the lowest tide occurs near midnight during the winter and near midday during the summer.[31]

Isolines, plotted at six-inch intervals, linking areas of equal tidal range, are shown on figure 11,4. As can be seen from this

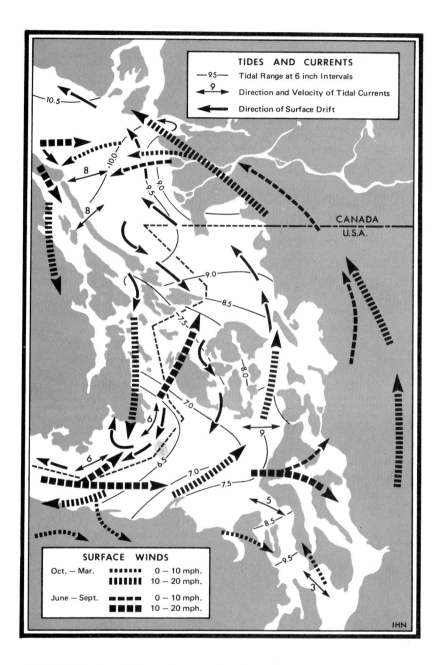

FIGURE 11,4   Tides and currents in Georgia and Juan de Fuca Straits.

illustration, the difference between mean low and high tides is relatively
small, 6.0 feet, in the Greater Victoria area, increasing to 7.5 feet at
the northern end of the Saanich Peninsula. These ranges increase to
more than 11 feet in the narrow channels to the north of Texada Island
and up to 15 feet near Olympia.[32] Tidal ranges in the area under
discussion are, therefore, relatively small, even the maximum range of
spring tides in the Saanich Peninsula rarely exceeds 10 feet. However,
it is during such periods of high tides, particularly when they coincide
with heavy precipitation and strong winds, that the bulk of coastal
erosion occurs.

(2) Currents

Water moving in response to tidal fluctuation across Juan de
Fuca and Georgia Straits becomes restricted by the Gulf Islands. Fast
tidal currents result, seen for example between Vancouver Island and
James Island. These vary from six to nine knots and are quite capable
of moving large quantities of eroded cliff materials.[33]

Similarly, currents are generated as a result of the addition of
fresh water to the Straits. By far the greatest source of such fresh water,
80 percent of that entering the Strait of Georgia, is provided by the
Fraser River. During the freshet season, late June and early July, this
river's discharge is 350,000 cubic feet per second. This fresh water
and that of lesser rivers, forms an upper layer above the saline ocean
water and flows seaward on its surface. Cold sea water, the northward
flow of which has been traced from the Oregon coast, moves at depth
through Juan de Fuca Strait.[34] Mixing of these two layers takes place
around the San Juan Islands and in narrow channels between other
islands, causing turbulence and setting up a density gradient which, in
turn, influences currents and water circulation in the Straits. Wind
induced currents also occur, but will be described elsewhere.

A systematic study of the currents of the Straits of Georgia and
Juan de Fuca has been in progress since 1968. A general counter-
clockwise rotation is suspected, as illustrated in figure 11,4. However, this is not permanent and can be reversed by fresh water runoff,
anomalous surface winds or tidal fluctuations.[35] The major significance
of such currents, to this study, is that they transport large quantities of
silt and clay, derived from eroding cliffs, into deep water, so preventing its accumulation along the eastern beaches of the Saanich Peninsula.

(3) Weather and Climate

The Saanich Peninsula and Greater Victoria areas enjoy a mild,
maritime climate. The Olympic Mountain Range in Washington State,
some 30 miles to the south, which reaches a maximum height of 7,913
feet in Mount Olympus, shelters southern Vancouver Island from the
major effects of many Pacific storms. Similarly, the hills to the west
of the Peninsula, rising some 3,000 feet in height, also give protection
from westerly, moist Pacific winds. As a result of its sheltered location,
the Saanich Peninsula lies in a rain shadow area, with Victoria receiving
only 27.1, Cordova Bay 32.1, and the Pat Bay Airport 33.6 inches of
annual precipitation.[36] Rainfall, however, increases westwards with
altitude, reaching a maximum of some 50 inches.

Of particular concern to any discussion of the impact of climate
and weather on rates of coastal erosion are the frequency and intensity
of severe storms, since these exercise a major influence on the wave
climate and, therefore, control both the maximum wave size and duration
of cliff attack.

W. J. Maunder classified over 700 surface daily weather maps,
for the years 1964 and 1965, which had been analyzed by the Department
of Transportation Weather Office. More than a dozen different types
of weather were found to have been significant to Vancouver Island

157

during this period. Of these 398 (54 percent) were classified as high pressure, 275 (38 percent) as low pressure and the remaining 58 (8 percent) as miscellaneous.[37] From this analysis, it appears that the area is normally under the influence of a mild westerly circulation which gives rise to warm weather in summer and considerable cloudiness and periods of rain in the winter. Northerly winds in summer result in hot clear weather but in winter are associated with cool, cloudy weather. Occasionally, cold Arctic air reaches the area accompanied by strong northeasterly wind and snow. The prevailing wind direction is from the north during the winter (October-March) and from the southwest during April to September. Winds of gale force from the southeast and southwest, preceding and following Pacific winter storms on the coast are also fairly common.[38] Wind speeds vary considerable, the maximum observed hourly speed at Gonzales Observatory (2 1/2 miles east-southeast of Victoria's city centre) has been 68 miles per hour, with gusts up to 90 miles per hour; while at the more sheltered International Airport (18 miles north of Victoria on the Saanich Peninsula) the maximum observed hourly speed was only 48 miles per hour, with gusts up to 52 miles per hour.[39] Naturally such winds have the greatest potential for stimulating erosion of those cliffs exposed to long distances of fetch.

Such erosion is particularly rapid during severe storms which drive high waves against the cliff face and may also cause surges. The occurrence of severe storms, defined as weather events lasting one or more consecutive days, with 4 inches or more precipitation on each day, has been analyzed for the lower Fraser Valley, by Sporns for the years 1921 to 1961 inclusively.[40] Many of these storms also affected the Saanich Peninsula and his discussion, therefore, is of value in analysing their impact. Sporns found that virtually all storms in the lower Fraser Valley occurred in the winter, being fairly evenly distributed from

October to February inclusively, with a slight maximum in December.
Severe storms were relatively rare during the rest of the year. The
total occurrence also varied greatly from year to year, six being
recorded in 1935 and none in 1948.[41] The number of days per year during
which storms occurred also varied, from a maximum of 38 in 1935 to none
in 1948.[42] Since wave height increases with wind velocity it is during
such weather conditions that large waves close the lower lying roads of
the city of Victoria to traffic, and rapidly erode coastal Pleistocene
sediments.[43]

The precipitation associated with large storms is also a major
factor affecting the rate of cliff erosion. Although direct runoff may
cause some gullying, by far the greatest effect of rainfall is its stimulus
of mass movement. Such collapse takes place when the shear stress to
which coastal sediments are exposed exceeds their shear strength. This
occurs when either the former increases or the latter decreases or when
both take place concurrently. Shear stress, for example, may be increased
by marine undercutting, waterlogging, the additional weight of new
buildings, transitory earthquake motion or the passing of heavy traffic.[44]
In the case of coastal erosion around the Saanich Peninsula the greatest
cause of increasing shear stress is the undercutting of unconsolidated
Pleistocene sediments by the sea during major storms. However, it
should be noted that this is also a high risk earthquake zone and large
scale coastal mass movements could occur during high magnitude seismic
events.[45]

Shear resistance is also decreased during storms because the shear
strength of clays, such as the Victoria and Post-Dashwood marine clays,
declines as their water content increases. The absorption of water may
lead to changes in the fabric of such clays while pore water pressure
also increases, further reducing shear resistance. Such weakening
processes also occur, in the clay rich Vashon tills, during periods of

FIGURE 12,4
Mass movement, Cowichan Head.

160

heavy rainfall.

Since increases in shear stress and declines in shear strength,
both occur during major storms, it is during these periods of heavy
rainfall that most cliff failures occur. Such movement is predominantly
slipping and occurs along a concave-upwards failure plane. Failure is
rotational along this slip surface and, as a result, induces a backward
tilt which can be seen from an inward-slopping head and heaving at
the toe.[46] Translational slides and falls also occur but are rarer. The
latter occur chiefly in summer when coarser grained surficial sediments
such as Vashon glacio-fluvial outwash and Quadra sands are dry and
are easily set into motion, in steep exposures, by strong winds.

When such failure takes place, cliff retreat occurs rapidly.
Fallen materials often extend far out onto the beach and are, therefore,
eroded at times when the sea does not normally reach the cliff face.
In this way, the debris is quickly removed and marine undercutting of
the cliff itself re-established, so increasing shear stress and initiating subsequent collapse. This erosional cycle functions most rapidly in winter
and is particularly destructive during major storms. Its operation is
generally limited to the thick surficial sediments of the south and east
of the Saanich Peninsula (Figure 12, 4).

(4) Beach materials

Beaches can be considered as 'rivers of sand', the movement of
which dissipates large quantities of marine energy. Where beaches are
extensive, they restrict the time available during which the sea can
erode the cliffs. In contrast, their absence allows smaller waves, during
lower tides, to erode the land margins. Destructive winter storm waves
decrease the volume of beach material which is built up once more by
more gentle constructive waves during fair-weather summer interludes.
A series of levelling exercises carried out from September 1967 to March

1968 at Island View Beach; September to December 1973 at Mount Douglas
Park Beach and September to December 1974 at Sayward Beach, established vertical changes in profiles of up to 6 feet. These correlated
well with wave and wind observations.[47]   This maximum depth of beach
disturbance, termed the sweep zone, is of great practical importance
since, when reduced in volume early in the Fall, beaches offer little
cliff protection should further storms then occur.

(5) Anthropogenic influences

      The effects of man-made perturbations on the marine-land
interface are often difficult to predict. Structures such as seawalls, like
that along Dallas Road, Victoria; groynes and breakwaters, such as that
almost linking Mary Tod Island to the Oak Bay mainland[48], have
immediate, if sometimes subtle effects on the rate of shoreline retreat.
(Figure 13,4). Their normal effect is to decrease longshore drifting,
causing oversteepening and erosion of the beach profile in the immediate
vicinity of the structure and slightly later, shoreline regression down
beach along barriers and spits. This removal of the supply of beach
sediment creates a local sink and an oversteepening of the longshore
gradient which returns to equilibrium by upbeach propagation of the
deficiency, back to the source headland. Here erosion is accelerated.[49]
Any attempt therefore, to alter coastal processes, including reducing the
rate of cliff retreat, effects other parts of the marine system. Such affects
may include the loss of recreational beaches and greater coastal erosion.
so increasing risk to property elsewhere.

RATES OF COASTAL EROSION

      It has been shown that numerous factors, both passive and
active, interact to influence rates of coastal erosion around the Saanich
Peninsula. In various spatial and temporal combinations these are

FIGURE 13,4
Breakwater, Oak Bay.

163

responsible for cliff retreat which varies from several feet per year to less than one inch in a century. Actual rates of marine encroachment have been determined using a variety of complimentary methods. The speed of cliff retreat at Cowichan Head, for example, has been measured by the author, continuously over a period of seven years (1968-1975). In other locations, erosion has been estimated after consulting archival materials, questioning long time cliff top residents and comparing vintage (pre-1945) and recent aerial photographs. Sections of coast undergoing varying rates of erosion are identified on figure 14,4, each is assigned a figure/number notation for ease of identification.

Rapid Erosion

These techniques have allowed the identification of five stretches of coastline which are retreating faster than one foot annually (Figure 14,4). Such erosion is classified as rapid and is commonly associated with high risk to life and, or, property. Each of these will now be described in detail, suggestions being made for appropriate future management strategies.

(1) Southwest of Esquimalt Lagoon (R1)

In this area, Vashon glacio-fluvial outwash sands and gravels of the Colwood 'delta' are under direct marine attack. Land wastage has been high, these cliffs have retreated 55 to 65 feet during the past 20 years, an average loss of 2 to 3 feet per annum. Total area losses during this time have been approximately 47,000 square yards, which would currently have a real estate value of $2 million.

Fortunately this area is not heavily populated and erosion has been associated with little structural loss. Legislation should ensure that this remains the case. In many respects, cliff retreat has been beneficial, since the eroded sediments have been moved northwards by

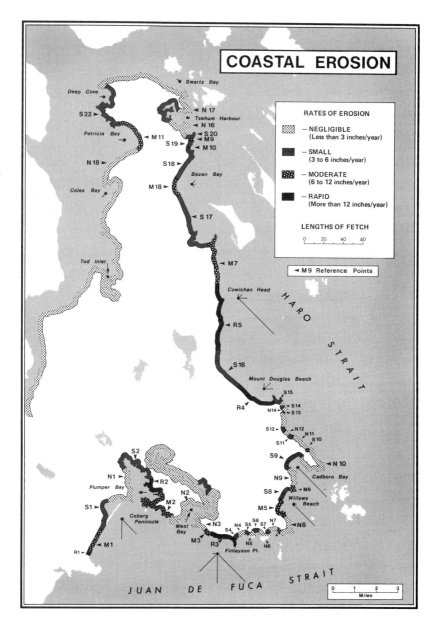

FIGURE 14,4   Rates of coastal erosion and lengths of fetch.

longshore drift, supplying and maintaining Esquimalt Spit. Some sand and gravel has also been carried southwards and, on deposition, has formed a second, smaller spit near Albert Head. Both spits have high recreational potential.

(2) Dyke Point to Cole Island (R2)

Pockets of Pleistocene till and glacio-fluvial material, in this district, are subject to mass movement when wet, particularly in areas of high relief. Those areas most affected have recorded losses of 2 to 3 feet per year during the past 35 years. This amounts to 53,000 square yards and would be worth $2.25 million at current market values. Unfortunately this high risk coastal area is heavily populated, with many of its residents living high above the ocean on bluff top locations. At the very least, careful consideration of the risks involved should be given before any further building permits are issued for such locations. Present residents should make every effort to maintain cliff face vegetation, even where this involves replanting the slip surfaces of mass movement phenomena.

(3) Holland Point to Clover Point (R3)

In this locality cliffs of thick Vashon till, overlain by Victoria clay, are exposed to storm waves from the south and the southwest (Figure 15,4). These unconsolidated materials are also prone to mass movement during periods of prolonged rainfall. In consequence, such cliffs remain very steep and are eroding rapidly. For example between Finlayson Point and Clover Point they are retreating at an average rate of some 14 inches annually. Since 1946 landward erosion has amounted to some 25 to 30 feet, a loss of 15,800 square yards.

Although such fairly rapid losses represent an undesirable reduction in the acreage of Beacon Hill Park, they are perhaps a necessary evil, since it is this erosion that supplies much of Victoria's beach sands

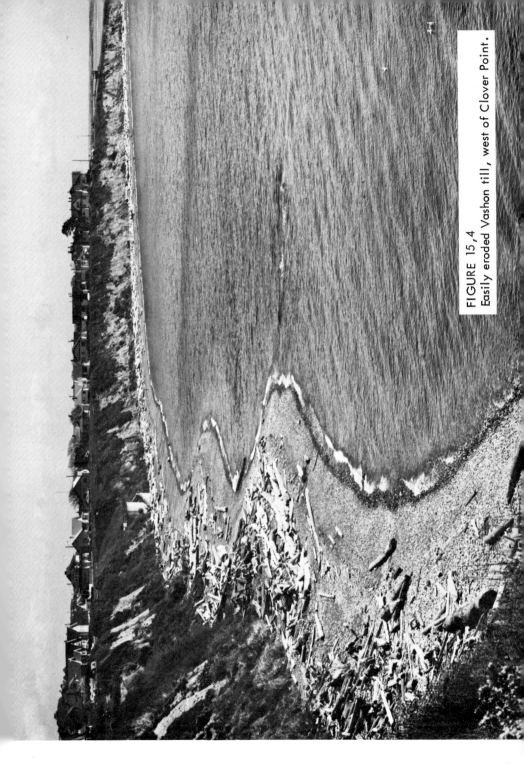

FIGURE 15,4
Easily eroded Vashon till, west of Clover Point.

and gravels. Fortunately, this entire area is parkland and cliff retreat presents, as yet, no threat to residential or commercial properties. The cliff top walkway will, however, require periodic relocation. Cliff mass movement is normally preceded, sometimes by many years, by the formation of a slip plane which is often visible on the ground surface as a semi-circular crack. Such cracks are now visible south of Douglas Street and Cook Street (Figure 16,4). At the former location, benches have been placed on the seaward side of this crack, on material which will inevitably fall into the sea. At the latter site, the stairway to the beach and part of the path is on the seaward side of the surface crack. These stairs, therefore, tend to encourage pedestrian usage of a location where large scale cliff failure is inevitable, and so greatly increase the risk of life loss. Their relocation would seem expedient.

(4) Cormorant Point to the north of Mount Douglas Park (R4)

The cliffs from Cormorant Point to Mount Douglas Park are composed of a very complex sequence of terrestrial, marine and glacio-fluvial sediments, some of which may even pre-date the Dashwood glaciation. These sediments have probably been severely cliffed in the relatively recent past, since they rise to heights approaching 250 feet in places. However, they are currently well vegetated and relatively stable, except where recent rotational slips or transitional slides have occurred.

Such slips are now undergoing some erosion but at slower rates than have probably occurred at various times in the recent past. To maintain what appears to be anomalously high stability, every effort must be made to encourage cliff face vegetation, including the replanting of failure surfaces. Surface drainage should not be allowed to enter the groundwater system. In this manner the relatively small risk to life and property may be prevented from increasing.

168

FIGURE 16,4
Coastal erosion threatening stairway and clifftop walkway.

The cliffs of Mount Douglas Park are composed predominantly of Victoria Clay and Vashon glacio-fluvial sands and gravels. They are exposed to wave attack from the north and northeast and have been re-treating rapidly, during the past 30 years, probably at rates of 3 to 4 feet per annum. This erosion represents an annual parkland loss of 600 square yards.

Saanich Council has recently attempted to reduce this wastage by commissioning the building of a seawall, composed chiefly of large boulders. Its main objective is the prevention of marine undercutting and associated rotational slips (Figure 17,4). Whether this manage-ment strategy will succeed remains in doubt. However, the wall is not sufficiently substantial to prevent marine erosion during major storms nor will it necessarily prevent cliff failure during periods of wet weather.

Whether such erosion should be prevented, even if this were feasible, can be questioned since, traditionally, the cliffs of Mount Douglas Park have nourished the beaches of Cordova Bay and possibly those as far south as Cormorant Point. Should this anti-erosion strategy prove successful, and coastal retreat be reduced or prevented at Mount Douglas Park, the sweep zones of these beaches may well be increased, their recreational quality may decline and residential properties in both areas may be subject to increased risk from accelerating coastal erosion.

(5) Cowichan Head (R5)

The cliffs at Cowichan Head rise to heights of some 200 feet and display a variety of Pleistocene sediments with a clarity unrivalled elsewhere in the area (Figure 8,4). Exposed in the cliffs are varying thicknesses of Post-Dashwood and Victoria marine clay, Quadra sands, Vashon glacio-fluvial outwash sands and gravels and fine, white, well-sorted sands of unknown (loessic?) origin. Rotational slips, often involv-ing many hundreds of cubic yards of material are common, particularly

FIGURE 17,4
Seawall, Mount Douglas Park, Saanich.

171

in winter. As a result, these cliffs are retreating some 5 to 6 feet
annually. In consequence, cliff top residents are suffering rapid property losses. This rate of erosion declines southwards to Sayward Beach
where losses average some 1.5 feet per year. Over this entire cliff
line at least 65,000 square yards of land have been lost during the past
20 years, an acreage which would have a current market value of
approximately $2,700,000.

This area experiences the highest risk from marine activity in
the Saanich Peninsula or the Greater Victoria areas. Despite this potential
for disaster, new housing is still under construction (August 1974) on
land which, if the current rate of coastal erosion continues, (and there
is no reason to suggest otherwise), will not exist at the expiration of a
normal mortgage. Indeed, some of the 'beach front' properties may be
in the sea within 10 years, should the cliff be attacked several times
by severe storm waves.

Cliff failure , in this area, generally occurs as a result of sand
runs in dry, windy weather and rotational slippage of water saturated
marine clays during the winter. An average rate for such losses is
misleading, since a large slip may remove 10 or 15 feet of the upper
ground surface in a single night, its fallen bulk then protecting the new
cliff face until it is finally removed by the sea.

The erosion of Cowichan Head can not be prevented without an
immense expenditure of public money. Such an investment would have
to be made by the Federal or Provincial governments and would be hard
to rationalize. The present policy of Central Saanich Municipality
which allows residential construction in this area is clearly irrational.
This error requires immediate rectification. Clearly no further construction should be permitted with 200 yards of this cliff line and the area
should be zoned for agricultural or recreational use. Thought should also

be given by Central Saanich Council and perhaps the Capital Regional District or the Provincial Government to the possibility of initiating some type of buy-back programme, which would allow residents to sell their property to the government.[50] At the very least, financial aid should be given to meet the cost of moving coastal houses. This process should not be long delayed since, as erosion continues and homes become located nearer to the cliff edge, the risk that the heavy vehicles and equipment required to move them may trigger mass movement and cause considerable life and property loss, increases. In the author's opinion, if nothing is done to mitigate the disaster potential in this area, extensive loss of property and possibly life can be anticipated within the next twenty years.

As the dangers become more apparent, demands for coastal sea defences in the Cowichan Head vicinity will probably grow. Even if it were possible to prevent erosion in this locality, it is not desirable since eroded material for these cliffs supplies both Island View Beach and Cordova Spit, both of which would be subsequently destroyed should Cowichan Head cease to erode. The destruction of the latter would, no doubt, quickly cause the inundation of the silted lagoonal area on its western side, so allowing the sea to attack a former, now abandoned cliff line, above which considerable new construction has recently taken place (Figure 18,4). In summary then, the protection of Cowichan Head, even if it were economically feasible, would almost certainly cause the loss of the most attractive recreational beach on the Saanich Peninsula, result in the destruction of agricultural land and create erosional problems for residential properties further north.

Island View Beach is currently subject to winter erosion which has threatened marine inundation of the low-lying agricultural land supported by former lagoonal sediments. Attempts to halt this process

173

FIGURE 18.4
Abandoned cliffs, Island View Beach.

have included probably illegal and certainly aesthetically unpleasing
bulldozing of the beach on at least one occasion. This action merely
increased the depth of the sweep zone, so causing accelerated erosion,
the process it was designed to prevent. Old car bodies have also been
used to prevent spit erosion with relatively little physical and no visual
success (Figures 19,4 and 20,4).

Moderate Erosion

Ten stretches of coastline , some 11.2 miles in total length, are
subject to retreat of between 6 to 12 inches usually. Such erosion is
considered moderate. Included in this category are most of the southern
coasts of Esquimalt (M2) and Oak Bay (M5), Island View Beach (M7),
Central Bazan Bay (M8) and much of Patricia Bay (M11) (see Figure 14,4).

Even this rate of coastal retreat can be a serious threat to property,
particularly where it is associated with large, temporally unpredictable
rotational slides. This is the case between Richards Island and Dyke
Point in Esquimalt (M2), where the rate of erosion is some 11 inches per
year. Here over 20 feet of coastline has been lost during the past 30
years, a process possibly accelerated by the construction of nearby Department of National Defence dock facilities.

In some areas, for example Bazan and Patricia Bays, although
erosion is occurring, since there are no large cliffs, risk to life is greatly
reduced. Small cliffs are also far easier to protect and there is considerable evidence, in the form of rock fill, wooden pilings and concrete
walls, to suggest that the residents of many of these areas are aware of
this option.

It is unrealistic to expect that land subject to moderate marine
attack, will remain underdeveloped. Certain precautions can, however,
reduce risk. Groyne systems should be avoided, since they may promote
beach depletion and hence accelerate coastal erosion elsewhere. The

FIGURE 19,4
An aid to spit retention, Island View Beach.

FIGURE 20,4
Car bodies "protecting" spit.

vegetation on cliff faces should also be retained whenever possible. Householders can protect their property by the addition of large blocks to its seaward side, so retarding erosion. However, if this technique is applied extensively, it will inevitably lead to declines in the quality of associated beaches, since it will shut off their sediment supply sources. Wherever possible, in such areas, new buildings should be set back at least 70 feet from the cliff.

Small and negligible erosion

Fortunately in the study area there are 83.8 miles of coast which suffer little, if any, erosion. Most of Esquimalt Harbour (N1 and S2), Gonzales (N6) and Ten Mile Point (N10), Tsehum Harbour (N16), the Southern side of Colburn Passage (N17) and the eastern shore of Saanich Inlet, south of Patricia Bay (N18), are included within this category. In these areas, coastal erosion presents no threat. However, where there is little marine circulation, the deposition of sediment may cause problems, as for example it does in Portage Inlet and the Gorge Water-way (Figure 21,4).

Such stable coasts generally occur where the Vancouver volcanics, Wark gneiss or Sannich granodiorite outcrop. These rocks are very resistant to erosion and are not, therefore, the source of materials for any of the areas' beaches. Ideally, social and economic considerations permitting, it is here that the large demand for coastal settlement and for the building of new marinas should be met. Although there are no coastal sites which can be built on with impunity, if constructed with care, such facilities should not cause undue disruption to sediment flows nor should their use subject inhabitants to high personal risk.

CONCLUSIONS

Development accelerates change and implies risk. Where

FIGURE 21,4
Recent sedimentation, Portage Inlet.

guided by informed controls, however, gains can be maximized and losses minimized. Such restrictive legislation should only be enacted where threats have been identified. Unfortunately, that portion of the total risk mosaic, represented by coastal erosion in the Saanich Peninsula and the Greater Victoria area, has never previously been adequately examined. As a result, many unnecessary risks, some of which may end in loss of life, and will certainly result in property destruction, have already been taken. It is hoped that this review of the problem, a small step in the necessary total risk mapping of the area, may encourage more rational future planning.

REFERENCES

1. The author would like to acknowledge the financial aid of
      National Research Council of Canada (A4378) and
      University of Victoria, Faculty Research Grant 08-652.
      Thanks are also extended to the many students, in-
      volved in geomorphology courses taught by the author
      at the University of Victoria, for aid in the field.
      In particular the efforts of Brian J. Deakin, Peter
      S. Furnell, S. McClellan and A. S. Harrison should
      be noted.

2. For detailed discussion of a wide variety of disasters see HEALY, R.
      J. Emergency Disaster Planning. New York: John Wiley,
      1969. Also of considerable value is HEWITT, K and
      BURTON, I. The Hazardousness of a Place. Toronto:
      University of Toronto Press, 1971.

3. See for example United States, Office of Emergency Preparedness,
      Disaster Preparedness. Washington: United States
      Government Printing Office, 1972.

4. United States, Office of Emergency Preparedness, Region Seven,
      Geologic Hazards and Public Problems. Washington:
      United States Government Printing Office, 1969.

5. This has been substantiated by disaster research in many distinct
      localities. For a wide ranging review see various articles
      in WHITE, G. F. (ed.) Natural Hazards. New York:
      Oxford University Press, 1974.

6. CHORLEY, R. J. "Models in Geomorphology," in CHORLEY, R. J.
      and HAGGETT, P, (eds.) Models in Geography. London:
      Methuen, 1967.

7. GRANT, D. R., LEWIS, C.F.M., MATHEWS, W. H., McDONALD,
      B. C. and SCOTT, J.S. "Coastal Geomorphology and
      Man," Background Paper prepared for the Coastal Zone
      Seminar, March 1972, Dartmouth, Nova Scotia later
      compiled for distribution by the Atlantic Unit, Water
      Management Service, Department of the Environment,
      Ottawa.

8. Ibid ., see also HAILS J.R. "A Review of Some Current Trends in Nearshore Research," Earth-Science Reviews, 10, (1974), pp. 171 - 202.

9. CLAPP, C.H. Geology of the Victoria and Saanich Map-areas, Vancouver Island, B. C. Canada Department of Mines Geological Survey, Memoir 36, Ottawa: Government Printing Office, 1913.

10. Ibid ., p.5 identifies them as dating from the lower Jurassic to the upper Triassic while MULLER, J.E. on his Geological Reconnaissance Map of Vancouver Island and Gulf Islands (open file map, 1971) Geological Survey of Canada, suggests they are Carboniferous to ? Devonian in age.

11. A brief, but informative guide to the geology of the Victoria area was prepared by J. T. Fyles for use on a field trip by the Productivity Committee, British Columbia Forest Service, February 18, 1971.

12. CARSON D.J.T. The Plutonic Rocks of Vancouver Island, British Columbia: Their Petrography, Chemistry, Age and Emplacement. Department of Energy, Mines and Resources, Paper 72 - 44. Ottawa: Information Canada, 1973, p. 9.

13. CLAPP, C.H.., op.cit ., pp. 55 - 77 and Fyles, J. T., p. 5.

14. Ibid.

15. WILKEN, E.B. "Landscape Parameters and Interpretation" in STANLEY - JONES,C.V. and BENSON, W.A. An Inventory of Land Resources and Resource Potentials in the Capital Regional District. Victoria: Capital Regional District Report, 1973, p. 109.

16. HOLLAND, S.S. Landforms of British Columbia: A Physiographic Outline British Columbia Department of Mines and Petroleum Resources, Bulletin 48. Victoria: Queen's Printer, 1964, pp. 37 - 38.

17. BROWN, E.H. The Relief and Drainage of Wales: A Study in Geomorphological Development. Cardiff: University of Wales, 1960, pp. 43 - 122.

18. ARMSTRONG, J.E., CRANDELL, D.R. EASTERBROOK, D.J. and NOBLE, J.B. "Late Pleistocene Stratigraphy and Chronology in Southwestern British Columbia and Northwestern Washington," Bull. Geol.Soc. Amer., 76 (1965), pp. 321 - 330.

19. Striations and grooves over the entire area have been mapped by the author and by students under his supervision. They demonstrate a predominantly southerly ice flow.

20. FLINT, R.F. Glacial and Pleistocene Geology. New York: John Wiley and Sons, 1967, pp. 302 - 364.

21. Deposits of two glaciations can be clearly identified in the cliffs of Cowichan Head, Central Saanich.

22. MATTHEWS, W.H., FYLES, J.G. and NASMITH, H.W. "Post-glacial Crustal Movements in Southwestern British Columbia and Adjacent Washington State," Canadian Journal of Earth Sciences, 7 (2), (1970), pp. 69 - 701. This reference deals only with the depth of ice in the Vashon glaciation but since Dashwood ice covered a comparable area its dimensions were probably similar.

23. EASTERBROOK, E.J. Principles of Geomorphology. New York: McGraw-Hill, 1969, p. 308.

24. MATHEWS, W.H., FYLES, J.G. and NASMITH, H.W., op.cit. pp. 690 - 702.

25. FOSTER, H.D."Geomorphology and Water Resource Management: Portage Inlet, A Case Study on Vancouver Island," The Canadian Geographer, XVI (2), (1971). pp. 128 - 143.

26. LAMPLUGH, G.W. "On Glacial Shell Beds in British Columbia," Quart, Jour. Geol.Soc. Lond., 42 (1886), pp. 203 - 230. See also DUFF, W. cited in MATTHEWS, W.H., FYLES, J.G. and NASMITH, H.W., op.cit., p. 697.

27. Ibid., p. 698.

28. DAY, J.H., FARSTAD, L. AND LAIRD, D.G. Soil Survey of Southeast Vancouver Island and Gulf Islands, British Columbia. British Columbia Soil Survey, Report 6, Ottawa: Queen's Printer.

29. Much of the timber on the beaches of Southern Vancouver Island is, of course, cut timber and has been accidently lost from log booms.

30. HOOS, L.M. and PACKMAN, G.A. The Fraser River Estuary, Status of Environmental Knowledge to 1974. Special Estuary Series No. I, Ottawa: Environment Canada, 1974.

31. MACKENZIE, R.J.D. Canadian Tide and Current Tables. 1975 Juan de Fuca and Georgia Straits, Vol. 5, Ottawa: Canadian Hydrographic Service, Marine Sciences Directorate, Department of the Environment, 1974.

32. BARKET, M.L. Water Resources and Related Land Uses Strait of Georgia-Puget Sound Basin, Geographical Paper No. 58, Ottawa: Lands Directorate, Environment Canada, 1974.

33. Ibid., p. 11.

34. DUXBURY, A.C. quoted in HEDSPETH, J.W. "Protection of Environmental Quality in Estuaries in GOLDMAN, C.R., McEVOY III, J. and RICHERSON, P.J. (eds.) in Environmental Quality and Water Development. San Francisco: Freeman, 1973, pp. 243 - 4.

35. WALDICHUK, M. "Physical Oceanography of the Strait of Georgia, British Columbia, " Journal of Fisheries Research Board of Canada, Vol. 24 (3), pp. 321 - 486.

36. This information has been taken from DAY, J.H., FARSTAD, L. and LAIRD, D.G., op. cit., p. 18.

37. MAUNDER W.J. "Synoptic Weather Patterns in the Pacific Northwest," Northwest Science, 32 (2), 1968, pp. 80 - 88.

38. Victoria City Weather Office, Annual Meteorological Summary 1963. Meteorological Branch, Department of Transport, 1963, p. 2.

39. Associate Committee on the National Building Code, Climatic Information for Building Design in Canada, 1960. National Research Council 11153. Ottawa: Queen's Printer, 1970. pp. i - 48. See also Climatic Normals: Vol. V, Wind. Ottawa: Meteorological Branch, Department of Transport, 551 - 552 (71), 1968, p. 11.

40. SPORNS, U. Occurrence of Severe Storms in the Lower Fraser Valley, B. C. Ottawa: Meteorological Branch, Department of Transport (CIR - 3631 REC - 404), 1962, pp. 1-11.

41. Ibid., p. 9.

42. Ibid., p. 10.

43. STRAHLER, A.N. Physical Geography. New York: John Wiley, 1969, pp. 171 - 172.

44. ECKEL, E.G. (ed.) Landslides and Engineering Practice. Highway Research Board, Special Report 29, Washington: NAS - NRC Publ. 544, 1958, pp. 1 - 323. See also COOKE, R.V. and DOORNKAMP, J.C. Geomorphology in Environmental Management: An Introduction. Oxford: Clarendon Press, 1974, pp. 128 - 166.

45. MILNE, W.G. and DAVENPORT, A.G. "Distribution of Earthquake Risk in Canada," Bulletin of the Seismological Society of America, 49, (1969), pp. 729 - 754.

46. COOKE, R.V. and DOORNKAMP, J.C. op. cit., pp. 144 - 145.

47. This work was completed as research projects by students taking Geography 376, (Geomorphology) and Geography 377 (Applied Geomorphology) under the author's supervision.

48. This breakwater has been the subject of considerable controversy, many Oak Bay residents blame it for a decline in beach quality. See BROWN, W. "'Disgusting Slime' laid to breakwater," The Daily Colonist, September 23, 1973, p. 28.

49. GRANT, D.R., LEWIS, C.F.M., MATHEWS, W.H., McDONALD B.C. and SCOTT, J.S. op.cit., pp. 184-187.

50. This type of financial assistance was given to the residences of Saint-Jean-Vianney by the Governments of Quebec and Canada when mass movement destroyed several houses and threatened many others.

CHAPTER 5

SEISMIC MICROZONATION
OF VICTORIA
A SOCIAL RESPONSE TO RISK

Vilho Wuorinen

University of Victoria

INTRODUCTION

Despite improving technology and increasing ability to buffer
against the harmful effects of nature, losses from natural disasters continue
to rise as both occupancy of hazardous areas and material wealth grow.
Dacy and Kunreuther have estimated that the average annual rate of increase in damage from hurricanes, floods, tornadoes, and earthquakes in
the United States for the period 1925 - 1965 was 2.5 percent per year.[2]
No equivalent figures are available for Canada, but based on a comparable standard of living and a faster rate of population growth, the same
conditions probably obtain in this county.

In North America, although damaging earthquakes occur less
frequently than floods, hurricanes, or tornadoes, the losses associated
with them have grown at a faster rate. In the United States, the rate of
escalation was 5.8 percent per year for the period 1925 - 1965, more than
double the average increase for the leading four disasters.[3] Of the natural hazards which might pose a catastrophic threat to Victoria, the city
is considered to be most vulnerable to earthquakes.

In man's quest to reduce harmful effects of natural hazards, he
must use all the means at this disposal. Seismic zoning maps and microzoning maps provide him with aids in making rational decisions about

adjustments to the earthquake hazard. The basic distinction between seismic zoning maps and microzoning maps is that the former portray an estimate of the risk of earthquake occurrence over a wide area, while the latter show local variation in the intensity with which an earthquake may be felt over a small area.

The first seismic zoning map for Canada was prepared in 1952 by Hodgson in cooperation with the Division of Building Research, National Research Council, and incorporated into the 1953 edition of the National Building Code.[4] Using a concept originally established in the United States, it divided the country into four zones of potential damage: Zone 3 indicated major damage, Zone 2 moderate damage, Zone 1 minor damage, and Zone 0 no damage. This map was based on a knowledge of the distribution of the larger earthquakes experienced in Canada in recorded times, as well as on general geological and geophysical considerations such as major fault zones.

A new seismic zoning map based on a computer analysis of all earthquakes recorded in Canada since 1899 was incorporated into the 1970 National Building Code.[5] The risk at any given location was based on the effects of all known earthquakes at that site, determined by calculating the maximum peak horizontal ground acceleration of each earthquake. For each site, the acceleration amplitude with a probability of annual exceedance of 1 in 100 was determined by statistical methods. The regions where maximum accelerations may be expected in southern Canada are the lower St. Lawrence River area and western British Columbia, including the whole of Vancouver Island.

More detailed local studies allow greater precision in assessing the seismic risk at a particular site. The most definitive work on earthquake risk in Canada was published by Milne and Davenport in 1969.[6] For western Canada, this study was based on a statistical computation of the 1479 earthquakes which occurred between 1899 and 1960. A peak

186

horizontal ground acceleration of 10.7 per cent gravity for a hundred year return period was assigned to Victoria in this study.[7] Based on an empirical relationship between intensity and acceleration, an intensity VIII earthquake may be derived for the city.[8] Since calculations in the study referred to firm soil conditions, even higher intensities may be anticipated in the least stable sedimentary areas.

As the acceleration data used in the new seismic zoning map were obtained by an averaging process, its results only apply to average ground conditions. In order to assess the relative potential for earthquake damage in a limited area, it is necessary to take into account differences in local geological conditions. This paper represents such an attempt to micro-zone the city of Victoria according to variations in earthquake risk.

EARTHQUAKE HISTORY OF VICTORIA

From the time of the earliest known seismic event recorded in 1841, the west coast of British Columbia has experienced one great earthquake (Richter magnitude 8.0) and five major earthquakes (Richter magnitude 7.0 to 7.9).[9] Earthquakes which produced intensities of VI or greater are shown on Table 1,5. Over one hundred earthquakes have been felt in the city in the period for which records are available.

Although no figures are available for the dollar losses incurred as a result of the June 23, 1946 earthquake, the damage was the most extensive recorded in the city's history. A maximum intensity of VII was reached, with a distinct areal pattern discernible in the intensities observed. Areas where the highest intensities were reached included the area around the intersection of Hillside Avenue and Quadra Street (160/87 on Figure 1,5), and the Fairfield district east from Cook Street along Chapman Street (151/85). In these areas walls cracked and many chimneys were knocked down. In other areas, such as the Rockland district (154/84), the earthquake was felt only as a slight vibration.

187

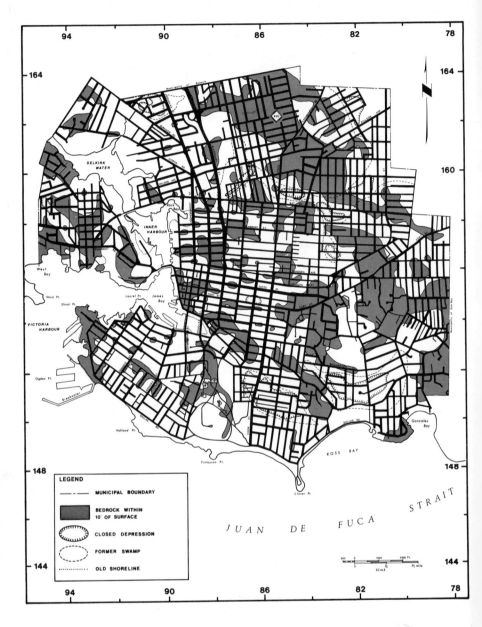

FIGURE 1,5   Location and depth to bedrock, Victoria.

VICTORIA EARTHQUAKES OF INTENSITY VI[9]

Date	Epicentre	Magnitude	Intensity
29 Oct. 1864	Gulf Islands	–	VI
25 Aug. 1865	Vancouver Island	–	VI
14 Dec. 1872	Chilliwack	7.5 (est.)	VI
16 Dec. 1872	Vancouver Island	–	VI
22 Aug. 1880	Gulf Islands	–	VI
22 Sept. 1891	Gulf Islands	–	VI
11 Jan. 1909	Gulf Islands	–	VI
6 Dec. 1918	Estevan Point	7.0	VII
12 Nov. 1939	Olympia, WA.	5.8	VI
23 Jun. 1946	North of Courtenay	7.3	VII
14 Feb. 1965	Puget Sound	6.5	VI

According to eyewitness reports, Victoria was a scene of wild disarray for a few moments on that day. Buildings swayed and plate glass windows bulged in and out as the tremor rocked the city. Telephone poles nodded to each other, with the wires whipping about like skipping ropes. Chimneys and steeples appeared ready to topple as they swayed, while hedges and sidewalks undulated in snake-like motion.

Occurring at 10:15 A.M. on Sunday, the earthquake caught most residents either at home or at church. There were no reports of physical injuries sustained by anyone although one woman fainted in St. Andrew's Cathedral (156/88).

Chimney breakage was the major damage most readily evident. Figure 2,5 shows one type of breakage, where part of the chimney rotated clockwise without toppling, others broke at the roofline and fell

to the ground. The Fire Department was kept busy all day pulling down
such damaged structures and found it prudent to warn all citizens against
the use of chimneys which might have been cracked unnoticeably inside
the house. They recommended immediate inspection of all flues to prevent serious fires from resulting.[10]

Only two public or commercial buildings were slightly damaged.
The old fire hall (157/88), which used to stand at the top of Broad Street,
in an area which is now part of Centennial Square, was cracked along
one wall to such an extent that girders had to be placed against it for
support. A new fire hall was constructed shortly after in a new location,
the move being hastened by the damage to the original building. Cracks
appeared in several places in the Empress Hotel (154/89), tiles were torn
from bathroom walls, and plaster fell in many parts of the old wing. The
most serious crack developed where the main building (completed in 1908)
and the south wing (completed in 1912) are joined (Figure 3,5).

GEOLOGY OF VICTORIA

Since geological and geomorphological conditions play an important part in the determination of earthquake hazard, all microzonation
methods are based on these considerations. Before the method used in this
study can be fully appreciated, the geology and geomorphology of the city
must be described in some detail.

Victoria lies in a glaciated lowland whose relief is broken by many
residual hills, three of which attain elevations slightly over sixty metres
above sea level. The extreme western portion of the city is disjoined by
a drowned valley.

Basement rocks are exposed in many parts of the city. The oldest
of these have been variously identified as being Devonian to Jurassic in
age.[11] During the Coast Range Orogeny, believed to have occurred

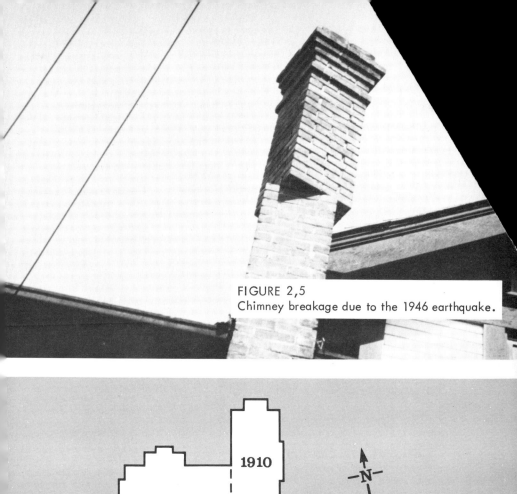

FIGURE 2,5
Chimney breakage due to the 1946 earthquake.

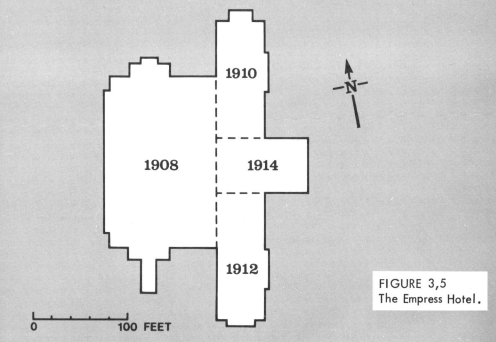

1910

1908

1914

1912

FIGURE 3,5
The Empress Hotel.

N

0   100 FEET

191

[12] these rocks were greatly deformed, metamorphosed, ...c plutons. Large folds, striking approximately ...ted by small folds and contortions. These rocks ...heared, and faulted.[13]

...has pointed out that the bedrock surface is very irregular ...city and the stratigraphy of the surficial deposits is very complex.[14] In the present study, examination of borehole records and visual inspection confirm both observations. Since the depth to bedrock and the type of surficial deposits are of particular interest in the prediction of seismic effects, both subjects are now considered more fully.

The bedrock surface in the coastal plain was eroded more or less into its present form during the early Pleistocene. Local depressions resulted from differential erosion along shear zones, joint planes and along the contacts between various rock formations. Modification of this surface by glaciation during the Pleistocene was probably limited to smoothing by the southward flowing ice.[15] This view is supported by the presence of unconsolidated sediments underlying drift at Cowichan Head, about 16 kilometres north of the city.

The extreme irregularity of the bedrock surface is confirmed by borehole records, street plan profiles, and visual examination of outcrops. The deepest borehole penetrated 60 metres before probable bedrock was encountered. The steep slopes of the surface are exemplified by the bedrock topography near the corner of Douglas Street and Kings Road (159/88), where bedrock is found 1 metre under Kings Road while 8 metres away, a borehole penetrated 18 metres of unconsolidated sediment without encountering bedrock.

Ground conditions affect earthquake damage potential in several ways. Primary effects may be considered to be the actual shaking of structures by seismic ground motion, while secondary effects are those phenomena initiated by the ground vibrations, such as settlement or liquefaction of cohesionless sediments.

Analysis of data from microseisms,[16] small earthquakes, and to a lesser degree large seismic events, have established that ground surface motions during earthquakes vary with ground conditions in a reasonably predictable way. Seed and Idriss found that in the four earthquakes, for which sufficient data were available on accelerations both in underlying rock and ground surface, the amplification factor of the unconsolidated deposits varied from 0.8 to 4.0.[17] Although the relationship has been clearly demonstrated, the research has suffered from a paucity of strong-motion seismograph data. As a result, in most cases practical assessment of the amplification factor has been based on empirical studies which have related general surficial sediment conditions to actual damage observed after earthquakes.

The ground vibration during an earthquake may compact loose sediments to such a degree that settlement is evident. During the 1964 Alaska earthquake, compaction of almost one metre in alluvium occurred at Homer.[18] Differential settlement of structures would be serious with this magnitude of compaction.

Liquefaction of saturated loose deposits has caused severe damage in many earthquakes. Seismic vibration compacts the surficial materials and increases the pore water pressure, causing the water in the voids to move upwards to the surface, as evidenced in mud spouts and sand boils. Structures may disappear into the resulting quicksand, while light underground structures such as septic tanks may be floated to the surface, as was the case in Niigata, Japan in 1964.[19] Again, differential settlement can result in extensive damage. Liquefaction in, or under, a sloping unconsolidated sediment mass can lead to flow slides. Even a thin layer of liquefied material under a surface of firm sediments may cause a serious slide over a wide area, as at Turnagain Heights, Anchorage, during the 1964 Alaska earthquake. Here, approximately 52 hectares

moved up to 180 metres laterally, destroying 75 houses.[20] The points
of most severe damage were at the toe of the slide and at the back where
houses dropped into the graben left by the surficial material as it moved
off.

Areas of loose deposits may fail without liquefaction occurring.
Seismic vibration may cause lateral spreading at the base of fill, leading
to cracking or collapse of structures built on it. The collapse of the
Sheffield Dam during the 1926 Santa Barbara earthquake is attributed to
such slumping of fill.[21]

Since various unconsolidated deposits react in different ways to
seismic vibration, the characteristics and distribution of these surficial
deposits should be considered in microzoning. The deposits of southern
Vancouver Island have been described by J. G. Fyles,[22] but have not
been discussed in detail for the Victoria area. An attempt has been
made here to correlate the local stratigraphy with Fyles' nomenclature
and to consider the seismic response of the various deposits by anology
to similar sediments elsewhere. The major stratigraphic units and their
correlative local deposits are listed in Table 2,5.

The oldest surficial materials found on Vancouver Island are the
non-glacial deposits termed Mapleguard Sediments by Fyles.[23] Although
small pockets of these sediments may exist in deep hollows in the bedrock
surface, they have not been identified locally.

Deposits of Dashwood drift can only be distinguished from the
younger Vashon drift by their stratigraphic position below the Quadra
Sediments. Clapp referred to this drift layer as Admiralty till and
described a one metre layer in southeastern Victoria.[24] Fyles considers
these deposits as being equivalent to Vashon drift, but notes other tills
in the Cordova Bay area, five kilometres north of the city, as possibly
correlating with the Dashwood.[25] No evidence of this drift layer can be

194

TABLE 2,5

VANCOUVER ISLAND STRATIGRAPHY

Major Unit	Local Deposit
Salish Sediments	Varied
Capilano Sediments	Marine sand, silt, Victoria Clay.
Vashon Drift	Vashon Till
Quadra Sediments	Cordova Sands and Gravels, Quadra Clay
Dashwood Drift	?
Mapleguard Sediments	?

After J. G. Fyles.

found in the borehole data.

The basal unit of the Quadra Sediments is a silty, clayey material which contains some sand, gravel, and occasional boulders. A layer of this Quadra Clay, over 15 metres thick, is exposed at Cowichan Head. It has not been identified in the borehole records, but it seems reasonable to assume that such deposits may be found in the deep depressions in the bedrock surface not reached during boring. Fyles notes that these clays are compact and stiff and able to support moderately large stresses without deformation.[26]

In the standard succession seen elsewhere on Vancouver Island, the Quadra Clay is overlain by a well-sorted deposit of fine - to - medium - grained sand, known locally as the Cordova (Quadra) Sands and Gravels. At higher elevations where younger deposits may have been removed by wave erosion, these sands may appear at the surface. Clapp suggested that thick deposits were protected from glacial erosion in the lee of the residual hills.[27] Throughout most of the city, however, the sand lies under Victoria Clay. This deposit represents a seismic hazard in that it might liquefy when saturated,[28] with similar results to those occurring in Niigata and Anchorage.

The Vashon till consists of gravel and boulders set in a dense silty, sandy matrix. Deposits up to 12 metres thick can be seen on the sea cliffs to the south of the city. Vashon till is compact and hard, and will not consolidate under heavy loads. When saturated, however, its high silt and clay content make it susceptible to sliding.[29]

Post-glacial materials, deposited when the relative sea level was at least six metres higher than at present, are termed Capilano Sediments on Vancouver Island.[30] Since few boreholes extend far beyond the rocks in the Vashon till, the records for the stratigraphy of the post-glacial deposits are more complete.

The lowest unit of the Capilano Sediments is a silty clay, known locally as Victoria Clay. By its stratigraphical position, this clay has been tentatively identified in 85 per cent of the boreholes for which data are available. Deposits reach 25 metres in thickness, with a desiccated upper layer evident in many cases. The oxidized upper layer is stable and not readily deformed under load. However, some of the unoxidized clays below the water-table may be quick clays.[31] Formed under marine conditions similar to those prevailing in the Ottawa-St. Lawrence River Valleys when the Leda Clay was deposited, it is not unreasonable to suspect that these clays may become unstable during a major seismic event.

Silty and sandy marine deposits cover some parts of the city, but usually only in thin layers seldom exceeding three metres in thickness. If such marine sands are saturated, they also might liquefy during seismic vibration.

196

Salish Sediments, related to present sea, river, and lake levels, do not appear to be extensive locally. Since artificial fills used to reclaim marine or marshy areas are usually less consolidated than natural deposits and, therefore, pose a much greater seismic risk, their distribution is of special interest in microzoning. The largest single area of fill is at the site of the Empress Hotel (154/89), where a six metre layer was laid in the reclamation of the head of James Bay.[32]

The complexity of the stratigraphy and the uneven distribution of the boreholes make it impossible to adequately map even the younger deposits of the Victoria area. However, several features which are of significance in seismic microzoning can be mapped with some accuracy and are shown on Figure 1,5. Areas where bedrock occurs within the top three metres are delineated. The pre-settlement drainage pattern is influenced both by the bedrock surface and the surficial deposits. In turn, the streams and swamps have slightly modified both the surface topography and sediments. Areas where extensive fill has been used to alter the shoreline are indicated on Figure 1,5.

Two major trends are obvious in the location of near-surface bedrock. An irregular ridge with minor interruptions stretches across central Victoria in a northwesterly direction. Roughly parallel to this feature is another irregular ridge extending from the eastern boundary to the northern boundary of the city. Despite their irregularity, these ridges appear to follow the line of major lineaments on Vancouver Island, for which the northwest strike of folding during the Coast Range Orogeny has been evoked as the cause. The second trend is the exposure of masses of bedrock on the residual hills, as might be expected. The largest of these hills is centered northwest of Government House (154/84), while Beacon Hill Park (152/88) is the centre of another. Smaller isolated pockets of bedrock occur throughout the city. Figure 4,5 a photograph of part of northeastern Victoria, illustrates the typical rocky terrain.

Extensive boggy areas existed before initial European settlement in northeastern Victoria; in the Fairfield area centred on Chapman Street (151/86); and between Government House and Ross Bay (151/83). Smaller swamps included one near the intersection of Vancouver Street and View Street (155/86) and another at the intersection of Vining Street and Stanley Street (157/83). Rithet's Bog (Figure 5,5) a present-day swamp about five kilometres north of Victoria, probably typifies the pre-settlement swamps of the city area.

As might be expected, peat and other highly organic soils and clays are found at or near the surface of the former swamps, as at the intersection of Linden Avenue and May Street (150/85). All boreholes in the former swamps show a clay layer, identified tentatively as Victoria Clay. Although the deepest borehole penetrates 23 metres of sediment, no rock has been encountered in any hole bored in former swampy areas.

The end of James Bay (154/89) is the site of the most extensive land reclamation in Victoria. Figure 6,5 shows the area before the turn of the century. The present-day causeway is located slightly further out towards the harbour than the Government Street bridge visible in the illustration. Another area of extensive fill occurs at Rock Bay (158/89).

It is clear from the preceding description of the irregular bedrock surface and the wide range of surficial sediments in the Victoria area that the degree of threat from seismic hazards can vary greatly even in locations in close proximity. The lack of depth to bedrock information over much of the city and uncertainty as to how local deposits would respond during a large earthquake make the construction of a microzoning map difficult. However, since earthquakes represent the most serious natural hazard in the area, it is essential that the production of such a map be attempted.

FIGURE 4,5
Housing on rocky terrain experiences
little earthquake damage.

FIGURE 5,5
Rithet's Bog.

199

METHODOLOGY

Various differing approaches to seismic microzonation have been developed during the past thirty years. Although all are based on geological and geophysical considerations, there is a considerable lack of uniformity in the ground characteristic parameters and techniques employed in determining them. Microzonation maps are in official use in the USSR, Japan, and New Zealand, while extensive research into the subject has been conducted in the United States and Chile.

By comparing local variations in intensity against ground types for thirty-four earthquakes in different Soviet cities, Medvedev developed seismic intensity increments for different kinds of ground.[33] The chief physical characteristic used in the evaluation of ground influence on this was calculated by multiplying the rate of propagation of longitudinal seismic waves in the top 10 metres by the density of that layer. Intensity increments ranged from 0.0 on granite to 3.9 on moist filled land or soil.

In 1954, Yokohama became the first Japanese city to be officially microzoned when the city was divided into three zones of varying ground conditions.[34] This ground classification and its corresponding design seismic coefficients had been drawn up on the basis of a comparison of geological conditions and damage experienced in previous earthquakes. The thickness of deposits is considered in addition to their characteristics. The depth to the water-table is accounted for only indirectly in the descriptions of the ground types.

In 1968, a Microzoning Committee of the Department of Scientific and Industrial Research was established in New Zealand, which was responsible in 1972 for the publication of a microzonation map for the city of Wellington.[35] Based on geology, soil type and geophysical information (gravity readings and microtremor measurements), the city was divided into three intensity zones and a landslide-prone area. Zone 1 comprised areas where bedrock was within 10 metres of the surface, Zone 2 included

FIGURE 6,5
View towards James Bay before
its reclamation.

B.C. Archive Photo

compact sediments more than 10 metres thick, and Zone 3 consisted of areas of highly porous sediments, including reclaimed land. The authors postulated that the amplification in vibration between successive zones would correspond to one step on the Modified Mercalli Scale.

Two salient features emerge from an examination of these and other microzonation maps. The record of previous earthquake intensities is the only quantitative control in assessing the anticipated variation in damage within a restricted area. Second, at present the most reliable method of mapping intensity zones on a micro-scale is the extension of the results of a correlation of this previous damage to the geologic conditions. These two features are common to most microzonation maps produced to date, and form the basis of the preliminary microzonation map of Victoria.

Research in the present study centred on gathering previously unrecorded data on the 23 June 1946 earthquake and correlation of this data with variations in local geomorphological conditions. Due to the paucity of recorded information, interviews were conducted with residents who had experienced that earthquake in Victoria.

In order to select respondents on a random basis, the city was divided into 72 squares. Of these, 16 had no dwellings in them, while 9 were only partially developed. Using random numbers to select intersections on grid squares, two locations per full square and one per partially developed square were determined, yielding a total of 103 locations. Two dwellings as close as possible to each of these locations were then selected by matching names of 1946 occupants with the current telephone directory, thereby providing a list of 206 possible respondents chosen at random. Two dwellings were selected at each location to enhance the probability of obtaining at least one response.

According to newspaper reports, damage was more severe in some localized areas of the city. In order to delineate these locations more accurately, as many potential respondents as possible were identified from these areas by the same procedure of comparing the 1946 city directory with the 1972 telephone directory. This added 211 possible contacts to the interview list, while 32 more were identified in the course of the interviews. Including 9 newspaper reports, a total of 458 discrete sources of information were investigated in this random stratified sample, representing approximately 3.4 per cent of the occupied dwellings in Victoria in 1946.[36] Rated responses were obtained in 202 cases.

The lower and upper limits of intensity on the Modified Mercalli Scale were found to be III and VII. Intensity I would not be reported, and the difference between II and III is impossible to assess in the case of a single report from one individual. The upper limit of VII was the intensity officially assigned to the impact of event in Victoria by seismologists at the time, and is corroborated by the evidence in this study.

Intensity can be assessed on the basis of an earthquake's effects on people or its impact on inanimate objects. Wherever possible, intensity was assigned in this study on the basis of effect on inanimate objects (for example, dishes broken rather than occupant frightened, for intensity V). In this connection, the quality of construction is an important criterion in differentiating intensities VI and VII. Damage to buildings was not used by itself in any assignment of intensity, so the question of deciding whether "poorly built" and "well built ordinary" buildings had been affected did not arise. At these intensities other criteria, such as broken chimneys and cracked or fallen plaster were used.

In twelve cases where other data were lacking, effects on personal reactions were the basis of intensity assessment at levels V and VI. In a study to determine how much weight should be given to personal reactions when they are unaccompanied by reports of corresponding effects on inanimate objects, Voigt and Byerly concluded that

the "effects on people" criteria justified raising the intensity from IV
to V or from V to VI if reports on nearby effects on inanimate objects
indicated the lower intensity.[37] The intensity assigned by "effects on
people" in this study is considered conservative, since the V or VI
ratings thus allocated were in areas where adjacent intensities based on
"inanimate objects" were equal or greater.

Information on the depth and stratigraphy of the surficial deposits
was gathered from a variety of sources. Data collected in the Urban
Geology Survey, conducted under the auspices of the Terrain Sciences
Division of the Geological Survey of Canada, are available on open
file at the Building Inspection Department, Capital Regional District
Planning Board. Results of 724 boreholes, distributed over 103 city
blocks, were available for Victoria. The coverage is uneven, with
comparatively high density in the downtown area where most of the
major construction has been undertaken.

To investigate the near-surface and surface areas of bedrock, the
area plans and plan profiles in the City Engineer's department were examined. The area plans show where bedrock was reached during trench-
ing for water mains, storm drains, gas lines, and underground telephone
or hydro lines. A plan profile exists for each street with any of these
services, and 453 profiles were examined to determine the depth of the
trench and the slope of the bedrock surface. The trenches varied from
less than one metre to eleven metres in depth, with an average depth of
three metres. All rock shown on the profiles could be confidently assumed to exist, whereas the possibility of rock occurring in cases where
it was not shown could not be ruled out. The data from these profiles
were plotted on the working map.

To complete the survey of rock outcrops, a visual examination
of every block in the city was conducted. In the downtown area most of

204

the outcrops have been blasted away and masked with pavement or buildings, but this is not so in the residential areas, thereby providing some balance for the paucity of borehole data from these areas. Tips of isolated boulders of similar lithology to local bedrock may have been confused with genuine outcrops during the fieldwork, but this problem resolved itself when the map was generalized at a later stage.

Preliminary studies had indicated that the old swampy areas and stream beds were of special significance in considering earthquake damage, so an effort was made to delineate the pre-settlement drainage pattern carefully. Three maps published in the period 1861 - 1884 and Clapp's 1910 map[38] show the location of swamps before they were drained, watercourses before their flow was diverted into the storm drain system, and the shoreline before it was modified by areas of extensive fill. The original drainage pattern is reflected in the present contours of land, and some of these features are shown on Figure 1,5.

The geomorphological data was plotted on a working map at a scale of 1:2400, generalized at a scale of 1:6000 before being reduced to a scale of 1:1200. Part of this city map is reproduced here as Figure 1,5 at further reduced scale.

Considered to be most accurately mapped are the areas where bedrock is at or within three metres of the surface. If errors exist here, they are errors of omission rather than ones of showing bedrock where it does not occur.

The contour interval on the working map was 1.5 metres, so the present-day closed depressions are accurately depicted. Only the ones related to old swamps are shown, however, since several closed depressions in the city are the vestiges of early quarries. The boundaries of pre-settlement swamps are based on the old maps which themselves varied considerably in accuracy. An attempt was made to correlate them with present contours, with the result that the boundaries shown reduce the area of the swamps indicated on the original maps.

Based on the local variation in response to the 1946 earthquake as determined from interviews and a comprehensive review of the stratigraphy and bedrock topography of Victoria, a preliminary microzonation map of the city was prepared.

THE MICROZONATION MAP

Earthquake wave motions are complex near the focus. Outside the epicentral area, this motion becomes even more complicated due to reflection and refraction at the boundaries of non-homogeneous layers in the earth's crust. Seismic wave motion and the resulting ground motion of the earth's surface at any one point will vary with the distance and bearing from the focus and with the magnitude, depth, and duration of the earthquake. It should be emphasized, therefore, that the present study is based on a particular earthquake, and the ground motion resulting from other earthquakes will differ.

A microzonation map based upon observed intensity has the major advantage that the effects of all factors are considered. The dominant influence of ground characteristics on local variations in intensity has been discussed earlier, but it should be stressed that the pattern of resulting damage may not be identical for each earthquake.

In considering the influence of ground characteristics it would have been better to test the correlation of intensity with depth to bedrock and type of surficial deposit independently. Lacking sufficient data to attempt this, the type of ground as differentiated on the surficial geology map was correlated with the intensities determined from the interviews. The results are shown in Table 3,5 and the Chi-square test of the independence of categorical variables indicated a positive correlation at a significance level of 0.01.

The lowest intensities were recorded on type A ground. It seems to make little difference what kind of surficial layers rest on bedrock if these layers are less than three metres thick. In most cases, foundations of structures reach bedrock, and in any event, the amplification of seismic waves is at a minimum in such thin layers.

Type C ground includes areas where fill has been used extensively in shoreline reclamation or where pre-development marshy ground was known to exist or postulated from a comparison of old maps and present-day topographic contours. As Table 3,5 shows, these areas correlated well with the highest intensities reached in the 1946 earthquake. How-ever, the cause of such higher ground motion amplitudes may not nece-ssarily have been simply the depth of fill or the type of deposit appearing near the surface. In discussing the greater damage incurred on old marsh sites in San Francisco in 1906, Evernden et al state:

> . . . there is a close correlation of deep stratigraphy and surface condition and it is not necessarily obvious that filling in the marsh was responsible for the high damage. The old marsh areas, as indicated by borings, are characterized by weak sedimentary types for appreciable thick-nesses below the surface. It seems much more likely to us that the old marsh sites were expressive of a general ground condition rather than that the ground conditions were caused by the marshes.[39]

Closely related to the areas of C type ground is the level of the water-table. The information available on the water-table is too incom-plete and variable in both time and location for a comprehensive picture of its nature in the city to be derived. However, data available indicate that the average depth to the water-table beneath areas of fill and former swamps is about three metres (Figure 6,5).

It is suggested that the higher intensities during the 1946 earth-

TABLE 3,5

CORRELATION OF INTENSITY/TYPE OF GROUND

| Type of Ground | Cases of measured intensity | | | | | |
	III	IV	V	VI	VII	Totals
A. Bedrock within 3 metres	29	30	3	0	0	62
B. Other than A or C	1	12	28	9	6	56
C. Fill or former swamp	0	0	10	52	22	84
Totals	30	42	61	61	28	202

quake in type C areas may be the result of higher amplification of
seismic waves in thick layers of saturated Victoria Clay. Before urban
development, such areas were commonly routes for surface drainage or
occupied by bogs, many of which are identifiable by the topographical
closed depressions now occupying the sites.

The ground classified as type B represents a great variety of
layers of differing thicknesses, and obviously includes small areas of
type A or type C ground which would have been so classified if more
complete data were available. For this reason, the range of intensities
found on this ground is wider than that of the other two categories.

Several conclusions may be drawn from the foregoing analysis.
There seems to be little doubt that areas where bedrock appears within
three metres of the surface would probably experience the lowest
seismic intensities in any future earthquake. In contrast, areas shown
on type C ground, known to be underlain by thick deposits of Victoria
Clay would probably experience the highest intensities. Intermediate
intensities are predicted for ground classified as type B, although a
wider range is possible. It is also concluded that a microzonation map
may be constructed with some confidence due to the excellent correlation
between intensity and ground type established in Table 3,5.

If a city were underlain by exceptionally homogeneous surficial deposits, no microzonation would be necessary and the intensity in any earthquake would be that shown for the area on a regional isoseismal map. At the other extreme, it might be feasible on a microzonation map to delineate as many zones as there were observed levels of intensity. In practice, the upper limit on the number of zones is predicated by other considerations. For engineering purposes, there is little advantage in differentiating areas where the expected intensity is V or less, since no structural damage occurs below this level. On the other hand, serious structural damage is general at intensity IX and little is gained in attempting to zone for higher intensities. In addition to giving the impression of greater sophistication in technique than is yet available, an excessive number of zones would militate against the basic purpose of the microzonation map in that the translation of the zonation into building code controls and land use planning would probably become overly cumbersome. Based on the variability of ground conditions in Victoria and the range of intensities observed in the 1946 earthquake, the city has been tentatively divided into three zones as defined in Table 4,5.

TABLE 4, 5

ZONES IN VICTORIA MICROZONATION

Zone	Related Ground Type	Intensity Increment
A	Bedrock within 3 metres	-1
B	Other than A or C	0
C	Fill or former swamp	+1

The median intensity reported in the 23 June 1946 earthquake was V, which corresponds with the mode for intensities on ground type B.

Since both the seismic intensity and ground characteristics of zone B
fall within the mid-range of values noted in this study, this zone is
postulated as representing the "average" seismic hazard in Victoria,
and therefore has an intensity increment of zero. However, it is emphasized that the wide range of intensities and the imprecise definition of
ground type in zone B lead to greater uncertainty about anticipated
intensities in this zone than in the others.

It has been possible to define the parameters for zones A and C
somewhat more firmly. The description for ground in zone A is quite
precise and three reports of intensity V were the only cases where the
mode intensity IV was exceeded in this zone. It is postulated that
intensities at least one unit lower than those in zone B would be experienced in zone A. Only ten intensity V observations of the total
eighty-four cases reported from zone C fell below the mode intensity
VI. Intensities of one unit or higher than zone B might be expected in
this zone.

Figure 7,5 is a reduced microzonation map. Close attention
was paid to present topography as this has an obvious relation to the underlying basement rock. Consideration was also given to the intensity ratings
determined from the interviews where such ratings provided refutation or
confirmation of postulated bedrock surfaces.

Since the resulting map closely resembles the surficial geology maps,
the same or somewhat more severe limitations of accuracy may be applied.
Zones A and C are considered to quite accurately delineated while a
slightly lesser degree of accuracy is claimed for zone B. Zone B is certain to contain small pockets of ground which belong to the other two
zones, and generally is the most variable ground.

The initial impression gained from an inspection of the microzonation map is that a relatively small proportion of the area falls into zone
C, while approximately half of the city lies in zone A with a comparatively

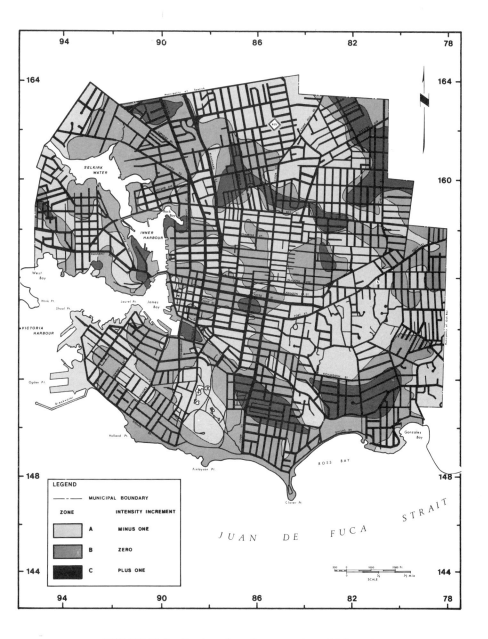

FIGURE 7,5   Earthquake microzonation of Victoria.

211

lower seismic hazard potential. In other words, if Victoria were to be included within the intensity VIII isoseismal as postulated for a hundred year return period, the zone C areas might be expected to reach intensity IX. On the other hand, more than half of the city might be expected to reach only intensity VII during the postulated earthquake.

Several points should be kept in mind when examining the map. The lines between zones represent intermediate bands where the intensity may be expected to vary within the limits of both contiguous zones. Zones A, B and C represent areas of differing anticipated intensities for an earthquake of any size which might be experienced, and bear no relation to the seismic risk zones discussed earlier. Any attempt to apply intensity increments as used here to modify risk zones would be meaningless.

The zones are based only on the differences in the amplification of seismic vibration by various surficial deposits, and secondary effects have not been considered in defining them. It would be wrong, however, to disregard the possibility of compaction, landslides, faulting or liquefaction occurring during a severe seismic event.

Victoria Clay is susceptible to compaction, as has been demonstrated by the differential settlement of the Empress Hotel. The south end of the structure has sunk 1.3 metres below the north end in sixty-five years. However, most of this compaction occurred in the first ten years following construction, and an examination of the time-settlement curves for the period 1912 - 1968 shows no abrupt increase in settlement in the years when the city experienced major earthquakes. This evidence suggests that sudden compaction from seismic vibration should not be ruled out as a possibility within ten years of the construction of a major structure on thick Victoria Clay. Where identified, these deposits are included in zone C due to their property of increasing the amplitude of seismic waves.

CONCLUSION

Tremors shake Victoria with great frequency, reminding citizens that a major earthquake is a distinct future possibility. It is not surprising, then, that interest in the subject continues to remain high. In discussing human adjustments to the seismic threat, attention has been focused on earthquake-resistant construction, since regulations governing the quality of construction are the only adjustment dictated by law at the present time.

In the 1970 National Building Code, one of the factors used in calculating the minimum lateral seismic force assumed to act on a structure is a foundation factor. This is to be 1.5 when the structure is founded on highly compressible soils and 1.0 for all other soils.[41] In terms of the present study, the factor to be used for zone C ground would be 1.5, for zone A ground 1.0, and either factor for zone B, depending on the density of the surficial sediments. The evidence presented here would seem to suggest that in addition to soil compressibility, the depth to bedrock should be considered in the definition of the parameters for the foundation factor. Pending future revision to these parameters, the present map could be used to ensure that a foundation factor of 1.5 is applied to construction in zone C.

The microzonation map may be most useful in planning future changes in land use. In this connection, two trends in the development of the city since 1946 should be mentioned. At that time, a few large areas of the city were underdeveloped. Several of the areas delimited as zone C, such as the area along Haultain Street east of Cook Street (159/84), were almost vacant in 1946, probably because they were unattractive for building as a result of poor drainage. By 1963, these areas had been fully developed as residential zones. A second trend is the expansion of multistorey buildings outside the downtown core. Experience in the Mexican earthquake of 28 July 1956 indicated that damage to multistorey buildings was more severe than damage to one or two storey buildings on poor ground.[42] The significance of these trends is that now much more of the poorest ground is developed than was the case in 1946, and some of this development is in the form of structures which are more susceptible to damage from strong ground motions. It might be noted that the two major proposed construction projects in the city in 1974, the Reid Centre on the waterfront between Yates Street and Fort Street (256/89), and the Victoria-Bapco development at Laurel Point (155/91), will both have their foundations in bedrock.

City zoning ordinances could be used to restrict certain types of development in zone C areas. These regulations might be in the form of restrictions on types of use, density of development, or maximum height of buildings. It is within the power of the municipal government to use tax incentives to promote development of zone A areas and discourage major building in zone C areas through differential assessment rates.

Another possibility in discouraging further development of zone C areas would be to ensure that buyers of property in such areas are fully informed of the comparatively high hazard in these locations. California provides this type of warning to buyers or lessees in subdivisions of more than five homes:

> The Bureau of Mines and Geology, State of California, reports that: This development lies within a fraction of a mile of the San Andreas Fault. In the event of a strong earthquake, severe ground movement, with attendant damage to the structures might be expected.[43]

Insurance against earthquake losses is available but is not carried by many property owners. This may be due to a lack of perception of the magnitude of the seismic hazards, or it may be that the cost of insur-

214

ance is considered excessive. On wood-frame construction, the
annual rate is five cents per hundred dollars coverage, compared to
sixteen cents per hundred dollars for fire insurance.[44] The rate varies
with type of construction, but not with ground characteristics. The microzonation map enables each Victorian to better assess the advisability
of buying such insurance. Even substantial buildings are liable to collapse in an intensity IX earthquake.

It is suggested that a preliminary microzonation map such as the
one discussed in this chapter should only be used for land-use planning
and as a basis for further investigation. Due to the limited data on which
it is based, under no circumstances could the map be considered a sub-
stitute for site geologic studies where required by the National Building
Code.

As a step toward eventual microzoning of urban areas, Canadian
scientists are investigating different techniques for determining local site
characteristics with the objective of predicting the variation in seismic
response.[45] This study represents an example of one such technique which
may contribute to the eventual microzonation of the Greater Victoria
Area.

REFERENCES

1. This paper is based on Wuorinen, V. "A Preliminary Seismic
    Microzonation of Victoria, British Columbia," (unpublished
    M. A. thesis, University of Victoria, 1974).

2. DACY, D. C. and KUNREUTHER, H. The Economics of Natural
    Disasters : Implications for Federal Policy. New York:
    The Free Press, 1969, p. 17.

3. Ibid.

4. WITHAM, K., MILNE, W. G. and SMITH, W. E. T., "The New
    Seismic Zoning Map for Canada: 1970 Edition, The
    Canadian Underwriter, (June 1970), p. 2.

5. Ibid., p. 4.

6. MILNE, W. G. and DAVENPORT, A. G. "Distribution of Earth-
    quake Risk in Canada", Bulletin of the Seismological
    Society of America, 59, (1969), pp. 729 - 54.

7. Ibid., p. 746.

8. Intensities noted throughout this paper are those defined in WOOD,
    H. D. and NEWMANN, F., "Modified Mercalli Intensity
    Scale of 1931", Bulletin of the Seismological Society of
    America, 21, (1931), pp. 178 - 82.

9. Source of these data is MILNE, W. G. "Seismic Activity in
    Canada West of the 113th Meridian, 1841 - 1951",
    Publications of the Dominion Observatory, 18, No. 7,
    (1964), and Victoria Daily Times, November 14, 1939, p. 1
    and April 29, 1965, p. 2. In some cases, intensities were
    estimated by the author from reports of damage.

10. Victoria Daily Times, June 24, 1946, p. 3.

11. J. E. MULLER dates it as Carboniferous to ? Devonian in his
    "Geological Reconnaissance Map of Vancouver Island
    and Gulf Islands" (open file map, 1971): CLAPP, C.H.
    places the age at lower Jurassic in Map 70A which is in-
    cluded in Geology of the Victoria and Saanich Map-Areas,
    Vancouver Island, B. C. Ottawa: Government Printing
    Bureau, 1913.

12. CARSON, D. J. T. The Plutonic Rocks of Vancouver Island, British Columbia: Their Petrography, Chemistry, Age and Emplacement. Ottawa: Information Canada, 1973, p. 9.

13. CLAPP, op. cit. , p. 5.

14. Ibid ., p. 25.

15. Ibid ., pp. 8 - 15.

16. Microseisms are not small earthquakes, but continuous disturbances recorded by seismographs. They are caused, inter alia, by machinery, traffic, wind, or breaking waves.

17. SEED, H. B. and IDRISS, I. M. "Influence of Soil Conditions on Ground Motions During Earthquakes," Journal of the Soil Mechanics and Foundations Division, American Society of Civil Engineers 95 (January 1969), p. 132.

18. SEED, H. B. "Soil Problems and Soil Behavior," in WIEGEL, R. L. (ed.) Earthquake Engineering. Englewood Cliffs, J. J.: Prentice Hall, 1970, p. 228.

19. Ibid ., p. 230

20. Ibid ., p. 239.

21. Ibid ., p. 243.

22. FYLES, J. G. Surficial Geology of Horne Lake and Parksville Map-Areas, Vancouver Island, British Columbia. Ottawa: Queen's Printer, 1963.

23. Ibid ., p. 15.

24. CLAPP, op. cit., p. 114.

25. FYLES, op. cit ., p. 19.

26. Ibid ., p. 102.

27. CLAPP, op. cit., p. 110.

28. FYLES, op. cit., p. 102.

29. Ibid.

30. Ibid., p. 74.

31. Ibid., p. 103.

32. CRAWFORD, C. B. and SUTHERLAND, J. G. "The Empress
Hotel, Victoria, British Columbia. Sixty-five Years of
Foundation Settlements," Canadian Geotechnical
Journal, 8 (1971), p. 82.

33. MEDVEDEV, S. V. Engineering Seismology. Jerusalem: Israel
Program for Scientific Translations, 1965, p. 38.

34. OHSAKI, Y. "Japanese Microzonation Methods." Paper
presented at the International Conference on
Microzonation for Safer Construction Research and
Application, Seattle, in November, 1972.

35. ADAMS, R. D. "Microzoning for Earthquake Effects in the
Wellington City Area," Bulletin of N. Z. Society
for Earthquake Engineering, 5 (1972), pp. 106 - 7.

36. No exact figures are available for the year 1946, but
there were 13,373 dwellings built in Victoria
before 1946. Canada, Dominion Bureau of Statistics,
1961 Census of Canada, Housing, Basic Dwelling
Characteristics, Bulletin 2.2 - 1, p. 18.1.

37. VOIGT, D. S. and BYERLY, P. "The Intensity of Earthquakes
as Rated from Questionnaires," The Bulletin of
the Seismological Society of America 39 (1949),
p. 26.

38. Map of Victoria and Part of Esquimalt Districts, London:
Day and Son, 1861; WADDINGTON, A. Map of Victoria,
San Francisco: C. C. Kuchel's, 1863; HARRIS, D. R.
A Map of the City of Victoria, n.p., 1884; CLAPP,
op. cit., map 20A.

39. EVERNDEN, J. F., HIBBARD, R. R. and SCHNEIDER, J. F.
"Interpretation of Seismic Intensity Data," Bulletin of
the Seismological Society of America 63 (1973), p. 420.

40. CRAWFORD AND SUTHERLAND, op. cit., pp. 77 - 93.

41. Associate Committee on the National Building Code. Canadian Structural Design Manual, 1970: Supplement No. 4 to the National Building Code of Canada. Ottawa: National Research Council, 1970, p. 14.

42. STEINBRUGGE, K. V. Earthquake Hazard in the San Francisco Bay Area: A Continuing Problem in Public Policy. Berkeley: University of California Printing Department, 1968, p. 30.

43. KATES, R. W. "Human Adjustment to Earthquake Hazard" in The Great Alaska Earthquake of 1964: Human Ecology. Washington: National Academy of Sciences, 1970, p. 26.

44. Rates are those quoted in 1974 for two per cent deductible by various local insurance companies.

45. MILNE, W. G. and ROGERS, G. C. "Evaluation of Earthquake Risks in Canada." Paper presented at the International Conference on Microzonation for Safer Construction Research and Application, Seattle, in November, 1972.

PLATE 5
Overview of Victoria's site.

CHAPTER 6

THE SIMULATION OF EARTHQUAKE DAMAGE

Harold D. Foster

and

R.F. Carey

University of Victoria

INTRODUCTION

The simulation of catastrophe is not an exercise in the macabre, but rather an essential first step in the mitigation of disaster.[1] All too often geophysical events cause numerous casualties and widespread damage because of community unpreparedness. Unfortunately, the citizens of Victoria suffer from this common lack of hazard awareness. Although the city lies in the highest seismic risk zone in Canada and has suffered several damaging earthquakes, a minimum of attention has been paid to the potential consequences of future shocks.[2] The following study, an attempt to simulate the probable impact of the 100-year earthquake, may help to eleviate this situation.

LITERATURE REVIEW

There is a surprising lack of published material dealing with the simulation of earthquakes and their impact on urban infrastructures. Much of the available literature, for example that of Whitman and his associates, deals with seismic impact on a single building of particular design.[3] The major concern of such civil engineering research is the construction, at minimum cost, of structures capable of withstanding maximum anticipated ground motion.

In contrast, very few attempts have been made to predict the total

impact of future seismic events on whole urban infrastructure. Exceptions
do exist, however, and include Friedman's estimates of damage potential
in the San Francisco Bay Area.[4] This particular study suffers greatly
because of the simplicity of its input data. Friedman, for example, did
not attempt to distinguish between frame or brick houses and made gen-
eralizations about land use which were extreme. Similarly, the study
by Cochrane, Haas and Kates of the future impact of a reoccurrence of
a 1906 - magnitude San Francisco earthquake, was based largely on
extremely generalized potential damage curves.[5] The following simulation of the effects of the 100-year earthquake on the city of Victoria
is unique in that it is based upon a building by building landuse survey.
As a result damage potential can be examined on a larger scale and in
more detail than has previously been possible.

SIMULATION TECHNIQUES

The simulation of anticipated earthquake damage in Victoria
involved five basic steps. Firstly, the average intensity of the 100-year
seismic event had to be established. The second necessary step was the
use of microzoning maps to predict spatial variations in anticipated intensity. Local seismic differences result from such varying factors as
the watertable depth, distance from the ground surface to the underlying
bedrock and the nature of the overlying surficial sediments.[6]

The social impact of an earthquake depends, to a large degree,
not only on the intensities involved but also the design and construction
of the buildings affected. Any simulation then should include, as a
third step, a detailed landuse survey. Every building is unique, yet
those of similar nature can be anticipated to respond analogously. The
fourth phase of the simulation then was that construction of a mean damage ratio matrix, an attempt to express the 'average damage' particular
building types have traditionally suffered as a result of ground motions

222

of known intensities.[7] The final stage of the modelling described here involved respresenting the distribution of damage potential on computer-printed maps. These maps were produced by comparing anticipated intensity (predicted from the microzonation of the city) with damage probabilities (derived from the mean damage ratio matrix). Each of these five stages will now be discussed in more detail.

Definition of Severity

The nature of earthquake risk in Canada has been described in detail by Whitham, Milne and Smith who, in 1970, published a seismic zoning map, based on an analysis of all earthquakes affecting the country since 1899.[8] The risk at any one location was derived from all known earthquakes which had influenced that site, and was calculated by determining the maximum peak horizontal ground acceleration of each seismic event. In this manner, Canada was subdivided into four zones, those areas where seismic risk was greatest being placed in Zone 3. Since Victoria is situated in this zone an acceleration of 6 percent gravity or greater having an annual probability of exceedance of 1 in 100 should be anticipated. This approximates to at least a Modified Mercalli Intensity of VII on 'normal' ground with intensities reaching as high as VIII in more vulnerable areas,[9] once a century.

This estimate contrasts with the earlier work of Milne and Davenport[16] who carried out an analysis of 1479 earthquakes influencing western Canada between 1899 and 1960. Their study postulated a peak horizontal ground acceleration of 10.7 percent gravity, for a hundred year return period; that is, approximately intensity VIII on 'normal' ground and IX on unstable sediments.[10]

223

Spatial Variations in Ground Motion

Wuorinen discussed those physical factors which influence the distribution of earthquake intensities, at the local level, in the preceding chapter. His microzonation of Victoria was used as input data for the simulation discussed here.[11] A few further qualifying comments appear necessary. Wuorinen's microzonation is limited to anticipated intensities associated with primary shock. Damage from aftershocks, and other hazards associated with earthquakes, such as sand boils and mass movement have not been included. Similar limitations, therefore, also apply to this study (Figure 2,6).

Structural Variations

Twenty-four significant categories of landuse and building type were established by a detailed survey of the literature of past seismic disasters.[12] These classes were based on response differences, noted during seismic ground motion experienced elsewhere. The classes identified are illustrated in Table 1,6.

Since no up-to-date information was available on land use in Victoria, a building by building survey was undertaken by the authors. To facilitate this, a grid system with dimensions of 99 by 127 was used. Each rectangle so produced represented an area of approximately forty by fifty-eight metres, within which one predominant land use was identified. Because of the large scale involved this was commonly a single building. Each structure or landuse was assigned to the appropriate category and is represented as such in Figure 1,6.

Estimating Probably Damage

While the accurate prediction of damage to individual buildings is extremely difficult to make, trends in structural response to seismic events

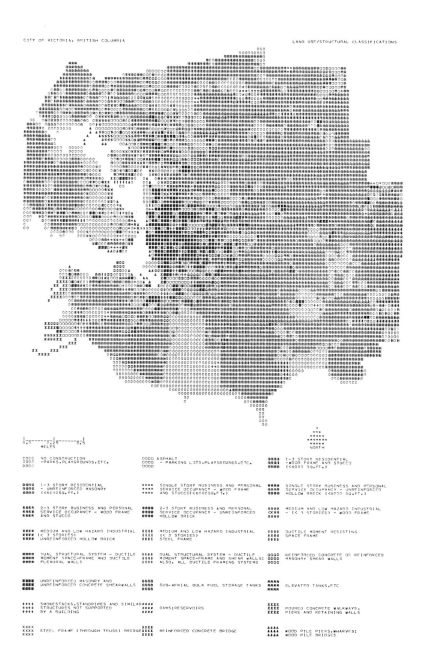

FIGURE 1,6   Landuse in Victoria.

225

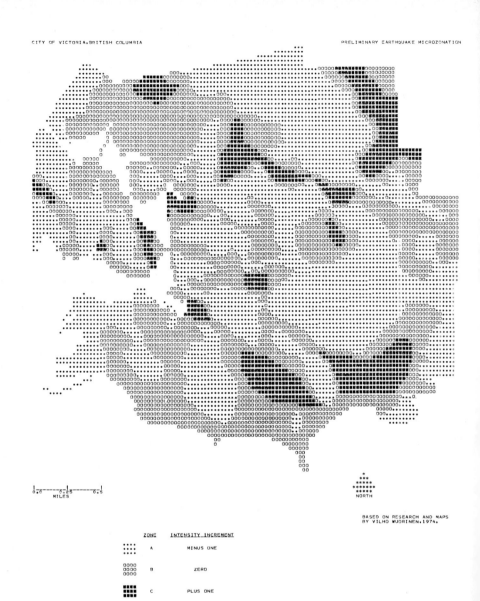

MILES

NORTH

BASED ON RESEARCH AND MAPS
BY VILHO WUORINEN, 1974.

ZONE	INTENSITY INCREMENT
A	MINUS ONE
B	ZERO
C	PLUS ONE

FIGURE 2,6   Computerized presentation of Figure 7,5 (after Wuorinen).

(by particular building types) have been identified elsewhere. These can be used to anticipate general patterns of damage associated with earthquakes of differing magnitudes.

The cost of repairing a building after an earthquake, expressed as a percentage of its cost of total replacement, is known as the damage ratio. A mean of these ratios, for all buildings within a particular category is known as the Mean Damage Ratio (MDR) and was developed by the authors for all twenty-four structural types used in the land use classification. This mean value replaces a full set of damage probabilities with a single 'average' figure.[13] Table 1,6 illustrates the resulting matrix. It consists of a Mean Damage Ratio for each structural classification, for earthquakes generating Modified Mercalli Intensities of VI to XI on 'normal' ground. Values one intensity above and one below the range actually simulated for Victoria were necessary because of the amplification of ground movement in areas of deep sediments (Zone C) and its reduction in areas where bedrock is at or near the surface (Zone A).

The information used to construct the mean damage ratio matrix, illustrated in Table 1,6 was derived from a variety of independent sources. For some structures, mean damage ratios have already been published, detailed damage statistics which allow their calculation have been made available for others by Steinbrugge.[14] In a few cases, only general descriptions of damage, caused by particular earthquakes to certain structures, were available. In this case, estimates of mean damage ratios were made using apparent trends in reactions of similar buildings for which more complete information had been published.

The damage that various structures can be expected to experience varies considerably, depending upon their design strategy, constructional material, location, and age. The quality of construction within a particular engineering system can also be a major factor.

Unreinforced masonry and concrete structures generally suffer greater

227

damage during earthquakes than wood-frame structures, primarily because
of their greater mass. Wood-frame dwellings tend to be more flexible
when subjected to vibration and have a high survival rate, even after
very large seismic shocks. Failure of wood-frame structures under four
stories during an earthquake is most often due to a lack of lateral force
bracing at foundation level, or poor condition of existing bracing, with a
resistant movement of the entire structure off its foundations. Modern
earthquake-resistant design strategies, and structures such as bridges or
fuel tanks tend to have mean damage ratios between these two extremes.[15]

Within the damage matrix, areas classified as 'no construction'
are shown to experience, of course, no structural damage. However, surface
upheaval, mass movement and other related phenomena can be expected at
intensities of VIII and above. This matrix (Table 1,6), showing the
'average' anticipated damage for particular earthquake intensities, together
with the land use survey and the microzonation map allowing prediction
of spatial distribution in intensities allow the simulation of earthquake
damage for the City of Victoria.

SPATIAL VARIATIONS IN DAMAGE POTENTIAL

Figures 3,6 to 5,6 printed directly by the computer line printer,
illustrate postulated seismic damage in Victoria, associated with earthquakes
capable of generating Modified Mercalli Intensities of between VII and X
on 'normal' ground (that is in Zone B). Such earthquakes would be
accompanied by intensities of between VIII and XI in Zone C (the least stable
ground) and VI to IX in Zone A (the most stable locations). It should be
noted that intensities of between VII to VIII may be anticipated once a
century in Zone B. The greater intensities are simulated for completeness,
but have very low probability of occurrence in any given year.

It should be noted once more that this simulation technique deals
with structural damage resulting from one primary seismic shock, aftershocks

or damage from mass movement phenomena are not considered. Similarly
secondary disasters, triggered by the initial shock, are not simulated.[18]
This is not to suggest that they could not occur. With severed power
lines, ruptured gas mains and fuel storage tanks, fire can quickly spread
into a conflagration following a major earthquake, as it did in 1906 in
San Francisco.[16] Disrupted water supply would severly hamper efforts
at control. The result may be substantially more destructive than the
initial earthquake. Damage from secondary causes is more difficult to
simulate because its distribution partially dependent upon factors, such
as the weather, industrial activities and population distribution at
particular times of day, which change constantly.

The major factors controlling damage during a seismic event are
its magnitude, distance of its epicentre from centres of population, the
nature of the substratum on which the affected settlement is built and the
type of structures and the mode of construction. The influence of many
of these variables can be clearly seen in Figure 3,6, which shows the
probable impact of an earthquake capable of generating intensities of
VII on 'normal' ground. This damage would approximate to the minimum
expected to be associated with the 100-year earthquake.

Over much of the city, little or no severe structural damage is
anticipated. Many exceptions, however, do appear to be probable. For
example, heavy between (20-65 per cent of replacement value) can be
expected to occur in close proximity to the Inner Harbour, east of
Government and Douglas Streets and in eastern Victoria West. The potential
for moderate damage (7.5 - 20 per cent of replacement value) is far more
widespread. It can, for example, be expected to occur throughout much
of the lower-lying parts of the Fairfield district and north of Ross Bay in
the south of the city and west of Douglas and north of Finlayson Streets,
in the other extremity. Numerous potential moderate damage pockets
can also be identified including the area west of City Hall. Many of
the one to three story wood-frame and stucco residential buildings in the

229

TABLE 1, 6
MEAN DAMAGE RATIO MATRIX

STRUCTURAL CLASSIFICATION	MODIFIED MERCALLI INTENSITY					
	VI	VII	VIII	IX	X	XI
No construction; parks, cemeteries.	0.0	0.0	0.0	0.0	0.0	0.0
Asphalt; playgrounds, tennis courts, parking lots.	1.0	15.0	40.0	66.0	100.0	100.0
One to three story residential; wood-frame; less than 6000 square feet.	0.3	1.25	8.25	12.0	20.0	50.0
One to three story residential; masonry; less than 6000 square feet.	2.0	4.0	20.0	70.0	98.0	100.0
Single story business and personal service occupancy; wood-frame; less than 6000 square feet.	0.5	1.5	9.0	15.0	25.0	60.0
Single story business and personal service occupancy; masonry and hollow brick, less than 6000 square feet.	1.0	3.0	20.0	75.0	99.0	100.0
Two to three story business and personal service occupancy; wood-frame; less than 6000 square feet.	1.0	2.5	12.0	22.0	35.0	75.0
Two to three story business and personal service occupancy; masonry and hollow brick; less than 6000 square feet.	2.0	5.0	25.0	85.0	100.0	100.0
Medium and low hazard industrial buildings; three stories or less; wood-frame.	2.0	4.0	20.0	92.0	100.0	100.0
Medium and low hazard industrial buildings; three stories or less; masonry or hollow brick.	2.25	4.5	22.5	99.0	100.0	100.0
Medium and low hazard industrial buildings; three stories or less; steel frame.	0.75	1.8	8.5	45.0	85.0	100.0
Buildings with a ductile moment resisting space frame.	4.0	8.5	18.0	45.0	65.0	85.0

TABLE 1,6 (cont'd)

MEAN DAMAGE RATIO MATRIX

STRUCTURAL CLASSIFICATION	MODIFIED MERCALLI INTENSITY					
	VI	VII	VIII	IX	X	XI
Buildings with a dual structural system consisting of a ductile moment resisting space frame and ductile flexural walls.	4.5	9.5	20.0	50.0	72.0	94.0
Buildings with a dual structural system consisting of a ductile moment resisting space frame and shear walls; also, buildings with ductile flexural walls.	5.0	10.5	22.0	54.0	80.0	100.0
Buildings with reinforced concrete shear walls.	0.13	1.4	10.0	45.0	95.0	100.0
Buildings with unreinforced masonry and unreinforced concrete frames and walls.	2.5	8.0	25.0	98.0	100.0	100.0
Sub-aerial bulk fuel storage tanks.	0.0	9.0	40.0	60.0	90.0	100.0
Storage tanks and contents other than FT; elevated tanks.	0.0	12.0	60.0	99.0	100.0	100.0
Smokestacks, standpipes, and similar structures not supported by a building.	20.0	60.0	100.0	100.0	100.0	100.0
Dams; reservoirs.	0.0	2.0	30.0	70.0	80.0	96.0
Poured concrete walkways, piers, and retaining walls.	0.0	2.0	25.0	65.0	75.0	90.0
Steel frame (through truss) bridges.	0.0	3.0	35.0	75.0	100.0	100.0
Reinforced concrete bridges.	0.0	8.0	40.0	80.0	100.0	100.0
Wood pile piers; woodpile wharves; wood pile bridges.	0.0	0.0	10.0	25.0	35.0	66.0

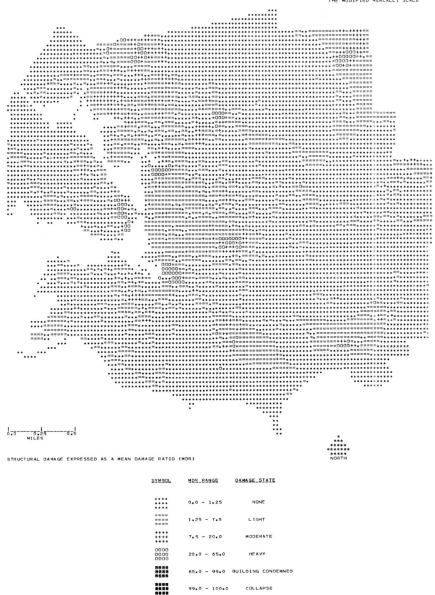

MILES

STRUCTURAL DAMAGE EXPRESSED AS A MEAN DAMAGE RATIO (MDR)

NORTH

SYMBOL	MDR RANGE	DAMAGE STATE
	0.0 - 1.25	NONE
	1.25 - 7.5	LIGHT
	7.5 - 20.0	MODERATE
	20.0 - 65.0	HEAVY
	65.0 - 99.0	BUILDING CONDEMNED
	99.0 - 100.0	COLLAPSE

FIGURE 3,6   Simulated earthquake damage in intensity VII.

city would however, suffer no appreciable damage in an earthquake generating such intensities.

Figure 4,6 illustrates the impact of an earthquake capable of intensity VIII on 'normal' ground. Naturally anticipated associated damage would be greater. Isolated examples of collapse (99-100 per cent of replacement value) could be expected. Similarly much of the Downtown area would suffer damage ranging from moderate to sufficient to make repair impossible or uneconomic. Other pockets of heavy damage might be expected in many locations, particularly at the western end of Johnson Street and south of Rock Bay. Although some buildings could be expected to suffer light or no damage, the city as a whole would be extremely hard hit by an earthquake capable of generating intensities of this size.

The damage associated with an earthquake capable of causing intensities of IX on 'normal' ground are shown in Figure 5,6. Several interesting patterns of damage are evident from this illustration. Again the greatest damage can be anticipated in the older sections of the city, generally east of the Gorge Waterway and west of Quadra Street. Exceptions do, however, occur and damage would also be extensive in Fairfield. In the unlikely event of an earthquake of this magnitude occuring in the near future, much of Downtown Victoria would suffer heavy or greater damage with numerous buildings collapsing.

CONCLUSIONS

Computer simulations of the damage potential of the one hundred year and greater earthquakes for the City of Victoria demonstrate several important factors. The older section of the city, including almost all the Central Business District is the highest risk area (Figure 6,6). Here, many nineteenth century brick and lime mortar industrial and business buildings are located on unstable sediments. Other, more modern buildings are

233

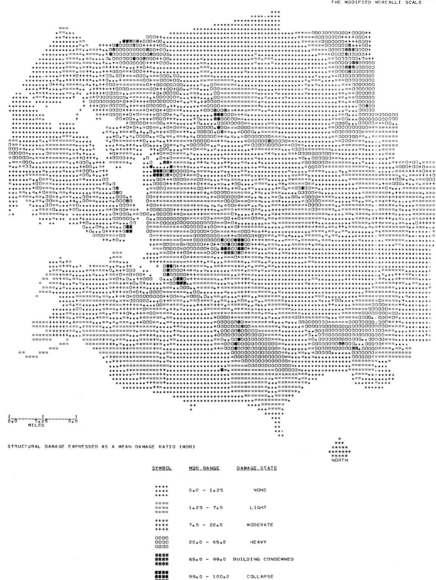

MILES

STRUCTURAL DAMAGE EXPRESSED AS A MEAN DAMAGE RATIO (MDR)

NORTH

SYMBOL	MDR RANGE	DAMAGE STATE
	0.0 - 1.25	NONE
	1.25 - 7.5	LIGHT
	7.5 - 20.0	MODERATE
	20.0 - 65.0	HEAVY
	65.0 - 99.0	BUILDING CONDEMNED
	99.0 - 100.0	COLLAPSE

FIGURE 4,6   Simulated earthquake damage in intensity VIII.

234

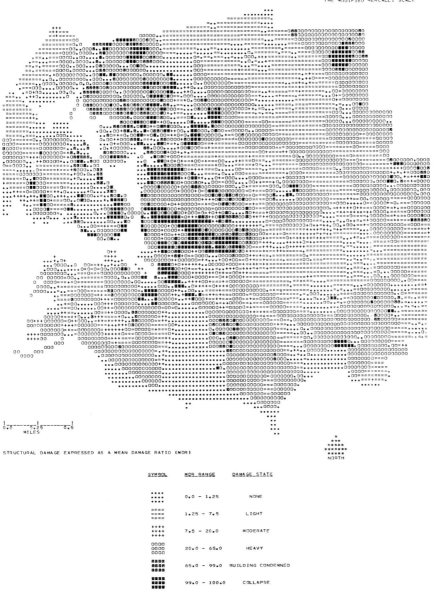

MILES

STRUCTURAL DAMAGE EXPRESSED AS A MEAN DAMAGE RATIO (MDR)

SYMBOL	MDR RANGE	DAMAGE STATE
∙∙∙∙	0.0 - 1.25	NONE
====	1.25 - 7.5	LIGHT
++++	7.5 - 20.0	MODERATE
0000	20.0 - 65.0	HEAVY
⧇⧇⧇⧇	65.0 - 99.0	BUILDING CONDEMNED
⬛⬛⬛⬛	99.0 - 100.0	COLLAPSE

FIGURE 5,6   Simulated earthquake damage in intensity IX.

235

FIGURE 6,6
High risk zone including the bus station.

often not earthquake-proof. Pockets of high risk also occur along former stream channels or where rock basins have been drained and used as construction sites (Figure 7,6). This situation contrasts with reasonably low risk in many residential areas, particularly where wooden frame buildings rest on or nearly on bedrock.

Such information should prove useful in planning the rational location of new buildings, particularly life-line facilities such as hospitals, emergency planning centres and communication networks. It also highlights the necessity for an up to date disaster plan which would accomodate such spatial differences in anticipated damage and locate evacuation routes and mobile medical and supervisory facilities accordingly. One further advantage of this simulation is that it allows a more rational calculation of earthquake insurance premiums and may provide a stimulus to the general public to purchase such coverage. Distasteful as disaster simulation may be to many, it must be remembered that ignorance of risk is never a guarantee against disaster.

FIGURE 7.6
Sharp contrasts in risk.

238

REFERENCES

1. See for example Office of Emergency Preparedness. Report to Congress: Disaster Preparedness. Washington: U.S. Printing Office, 1972.

2. The City of Victoria has, for example, no disaster plan which outlines how to accomodate the deaths, injuries and damage anticipated in the 100-year earthquake.

3. WHITMAN, R.V. "Damage Probability Matrices for Prototype Buildings," Seismic Design Decision Analysis: Report No. 8 (Structures Publication No. 380) Cambridge, Massachusetts: M.I.T., 1973.

4. FRIEDMAN, D.G. "Computer Simulation of the Earthquake Hazard" in OLSON, R.A. and WALLACE, M.M. Geologic Hazards and Public Problems. Santa Rosa: U.S. Government Printing Office, 1969, pp .153-181

5. COCHRANE, H.C., HAAS, J.E., BOWDEN, M.J. and KATES, R.W. Social Science Perspectives on the Coming San Francisco Earthquake Economic Impact, Prediction and Reconstruction. (Natural Hazard Working Paper No. 25). Boulder: University of Boulder, 1974, pp. 22-32.

6. See WUORINEN, V. Chapter 5 in this volume.

7. Mean damage ratios for some structural types were available in WHITMAN, op. cit. and in WHITMAN, R.V., BIGGS, J.M., BRENNAN, J., CORNELL, C.A., NEUFVILLE, R and VANMARCKE, E.H. "Summary of Methodology and Pilot Application," Seismic Design Decision Analysis: Report No. 9 (Structures Publication No. 381) Cambridge Massachusetts: M.I.T., 1973.

8. WHITHAM, K., MILNE, W.A. and SMITH, W.E.T. "The New Seismic Zoning Map for Canada: 1970 Edition," The Canadian Underwriter, (1970), pp. 1-9.

9. The more vulnerable areas are, of course, those where the bedrock is deep below the surface. See WUORINEN, Chapter 6.

10. MILNE, W.A. and DAVENPORT, A.G., "Distribution of Earthquake Risk in Canada," Bulletin of the Seismological

Society of America, 59 (1960), p. 739.

11.   See WUORINEN, V., A Preliminary Seismic Microzonation of
       Victoria, British Columbia. Unpublished M.A. Thesis,
       University of Victoria, 1974.

12.   Detailed statistics of structural damage caused by earthquakes are
       available from STEINBRUGGE, K.V. and BUSH, V.R.
       "Review of Earthquake Damage in the Western United
       States, 1931-1964" in CARDER, D.S. (ed.) Earthquake
       Investigations in the Western United States. Washington:
       U.S. Department of Commerce, 1965, pp. 223-256
       and in STEINBRUGGE, K.V., "Earthquake Damage and
       Structural Performance in the United States," in WIEGEL,R.L.
       (ed.) Earthquake Engineering. New Jersey: Prentice Hall,
       1970, pp. 167-226. Also of value was BERG, G.V.,
       "Response of Buildings in Anchorage," in Committee on the
       Alaska Earthquake. The Great Alaska Earthquake of
       1964: Engineering. Washington: National Academy of
       Sciences, 1970, pp. 247-282. Numerous other references
       were also consulted.

13.   When only a damage state was known for a particular structural
       type, MDR was evaluated as

$$MDR_I = \sum_{DS} P_{DSI} \cdot CDR_{DS} = \sum_{DS} \frac{n_{DSI}}{n_I} \cdot CDR_{DS}$$

       where P_{DSI} = probability that a building in a particular
                        category experiences damage state DS when
                        subjected to ground motion intensity I.
              n_{DSI} = number of buildings in a particular category
                        experiencing damage state DS when subjected
                        to ground motion intensity I.

14.   These references have been described in footnote 12.

15.   Ibid.

16.   HODGSON, J.H. Earthquakes and Earth Structure. Englewood
       Cliffes, New Jersey: Prentice Hall, 1964, pp. 10-16.

CHAPTER 7

GEOMORPHIC ASPECTS OF SANITARY LANDFILL SITE SELECTION

Terry Fenge

University of Victoria

INTRODUCTION

Municipal solid waste disposal (refuse) has become a striking environmental problem in recent years. Population growth, rising material living standards, and the 'throw away' mentality of urbanized, environmentally alienated consumers cause huge waste piles, expensive to handle and difficult to 'get rid of'. The U.S.A. is the world's largest waste producer, 170 million tons generated in 1965. By 1980 it is estimated that its waste will exceed 340 million tons.[1] In per capita terms the 'average American' created 4.5 lbs of household refuse per day in 1965, by 1980 it will probably be 8.0 lbs.[2] Canadians are the world's second largest per capita waste producers, a 1970 estimate being nearly 6 lbs per day.[3]

Refuse make-up varies temporally and spatially, reflecting life style, affluence, and attitudes of the discarding community (Table 1, 7). Adoption of the convenience principle in product marketing, consumer acceptance of disposable and rejection of returnable containers, has increased paper, paper products, plastics and glass used and subsequently thrown away.[4] The production of grass clippings, vegetables and ashes vary seasonally as well as regionally. Although food wastes (garbage) is declining as a percentage of total refuse it remains significant, being strongly correlated to income levels. People with low disposable incomes do not waste food !

Sophisticated technology, the assumed panacea to so many environmental ills, is strangely absent from the solid waste scene, for like

TABLE 1,7

COMPOSITION OF URBAN SOLID WASTE

(per cent composition of refuse)

Average Individual's Waste Production*		Hartland Road Sanitary Landfill**	
Per cent		Per cent	
23.38	corrugated paper boxes	19.0	wood
9.40	newspapers	12.0	metal
6.80	magazine paper	40.0	paper
5.57	brown paper	10.0	organic, food
2.75	mail		wastes
2.06	paper food cartons	5.0	glass
1.98	tissue paper	5.0	plastics
0.76	plastic coated paper	4.0	bricks
0.76	wax cartons	2.0	concrete
2.29	vegetable food wastes	3.0	rags
1.53	citrus rinds and seeds	100.0	
2.29	meat scraps, cooked		
2.29	fried fats		
2.29	wood		
2.29	ripe tree leaves		
1.53	flower garden plants		
1.53	lawn grass, green		
1.53	evergreens		
0.76	plastics		
0.76	rags		
0.38	leather goods		
0.38	rubber composition		
0.76	paints and oils		
0.76	vacuum cleaner catch		
1.53	dirt		
6.86	metals		
7.73	glass, ceramics, ash		
9.05	adjusted moisture		
100.0			

Sources: * Pollution of Groundwater Due to Municipal Dumps, by Hughes, G. et al, Inland Waters Branch, Department of Energy, Mines, and Resources. Technical Bulletin, No. 42 (1971) p.2.
**Evaluation of Groundwater Pollution at Hartland Road Sanitary Landfill, by B. and H. Levelton and Associates Ltd., September, 1971, p.3. An estimate of refuse make up received.

Neolithic man we still dump or bury the vast majority of our household wastes. Although recycling is gaining in popularity with the increasing realization of the finite nature of many resources, still over 80 per cent of municipal solid waste is disposed to land by open dump or its cosmeticised cousin the sanitary landfill.[5] Once dumped, refuse loses utility, it ceases to be a resource and reverts to a potential pollutant, defined by the National Academy of Sciences as a 'resource out of place'.[6]

In North America, with political pressure from central government, the sanitary landfill is becoming the disposal norm.[7] It was originally developed in England in the 1920's where it was termed 'controlled tipping'. This process is defined by the American Society of Civil Engineers as:

> A method of disposing of refuse on land without creating a nuisance or a hazard to public health or safety, by utilizing the principles of engineering to confine the refuse to the smallest practical area, to reduce it to the smallest practical volume and to cover it with a layer of earth at the conclusion of each day's operations, or at such more frequent intervals as may be necessary.[8]

Sanitary landfills are in use throughout North America, with slight modifications their applicability ranges from permafrost zones in the Canadian north to humid sub-tropical areas in southern Florida.[9,10] Two types are recognized.[11] On low, flat land, area landfills utilize the trench or ramp method; refuse is deposited in excavated trenches six feet deep fifteen to fifty feet wide and perhaps five hundred feet long. Excavated earth is used for cover. Depression landfills use natural or man made 'holes'; for example; quarries, gravel pits and canyons; cover material is often trucked in. Operations are highly mechanised yet simple. Average disposal costs are 50c–$4 per ton, far cheaper than such sophisticated technologic disposal as incineration or composting.[12] Yet increased waste production and additional landfills, have prompted scientist and public alike to perceive the environmental drawbacks despite the topics mundanity

(Table 2,7). Since passage of the 1965 Solid Waste Disposal Act and the 1970 National Environmental Policy Act in the United States, much research has focussed on these pollution side effects, perhaps reflecting governmental belief that sanitary landfill is likely to dominate waste disposal for many years. The best documented and potentially most serious problem associated with the method is leachate contamination of ground and/or surface water (Table 3,7). Precipitation and runoff percolate through cover material, entering decomposing refuse, leaching out soluble substances and degradation products, together with suspende solids.[13] This grossly polluted liquid emerges at the surface, if the fill is underlaid by impermeable clays, or filters through surficial material, reaching and then travelling with groundwater. Studies show leachate to contain high concentrations of an array of substances, its basic nature reflecting whether the decompositional environment is aerobic or anaerobic. Leachate from the former has a lower organic content (as measured by Biological Oxygen Demand or the Chemical Oxygen Demand), its decompositional end products are more highly oxidized, for example nitrates, sulphates, and ferric salts, it is also lighter in color with less odour. Anaerobic leachate contains mostly reduced compound and ions; such as sulphides, ammonium and ferrous salts. It has high B.O.D. or C.O.D. because of organic acids in the liquid and a strong musty odour due to partly degraded proteinaceous compounds,[14] (Table Definitive work undertaken at the Riverside landfill in California indicated, in the first year after dumping:

> It may be expected that continuous leaching
> of an acre foot of sanitary landfill will result
> in a minimum (emphasis added) extraction of
> approximately 1.5 tons of sodium plus potassium,
> 1.0 tons of calcium plus magnesium, 0.91 tons of
> chloride, and 0.23 tons of bicarbonate. Re-
> moval of these quantities would continue with
> subsequent years, but at a very slow rate. It
> is unlikely that all ions would ever be removed."[15]

244

TABLE 2,7

SUMMARY OF ADVERSE IMPACTS AND COUNTERACTIVE MEASURES
ASSOCIATED WITH SANITARY LANDFILLING

Anticipated Adverse Impact	Actions Planned to Mitigate Adverse Impact
Public Health and Aesthics	
Litter	Provide proper fencing. Control working face area.
Dust	Periodic watering
Odors	Assure prompt and consistent coverage of exposed wastes.
Leachates	Diversion of runoff and drainage of precipitation incipient on the surface. If necessary, install underdrains and a collection/treatment system.
Air Quality Impairment	Control dusts.
Heavy Equipment and Collection Vehicle Movement	Provide proper traffic directors and spatters. Assure adequate access roads.
Methane Gas Generation	Install appropriate gas control venting system. Minimize water infiltration to waste by drainage control.
Local and Regional Biota	
Vegetation	Remove only the vegetation necessary for operations. Install gas vents to preclude root-zone damage to adjacent vegetation.
Animal Life	Landscape finished landfill to reattract displaced native species. Control leachates from entering water courses.
Land and Land Use	
Visual Unattractiveness	Plan cut and fill areas to minimize "desecration" appearances.
Restricted Land Use	Plan for parks, golf courses, and open space.
Social and Economic Environments	
Public Opposition	Develop a comprehensive public relations/education program to promote and explain need for sanitary landfill and its operation. Arrange for public meetings to air grievances-dispel aura of public powerlessness and promote participation in planning process.
Cost Increase	Incorporate discussions for landfill economics into public relations programme.

Source: STEARNS, R.P. and ROSS, D.E., "Environmental Impact Statements for
Sanitary Landfills," Public Works (Nov. 1973), p. 65.

245

TABLE 3,7

LEACHATE COMPOSITION

Characteristic (mg/l)	Source					
	1+	2+	3+	4+	5+	6+
pH	5.6	5.9	8.3	–	–	–
Total hardness as $CaCO_3$	8,120	3,260	537	–	–	500
Iron total	305	336	219	1,000	8,700	–
Sodium	1,805	350	600	–	–	–
Potassium	1,860	655	N.R.	–	–	–
Sulfate	630	1,220	99	–	940	24
Chloride	2,240	N.R.	300	2,000	1,000	220
Nitrate	N.R.	5	18	–	–	–
Alkalinity as $CaCO_3$	8,100	1,710	1,290	–	–	–
Ammonia nitrogen	845	141	N.R.	–	–	–
Organic nitrogen	550	152	N.R.	–	–	–
COD	N.R.	7,130	N.R.	750,000	–	–
BOD	32,400	7,050	N.R.	720,000	–	–
Total dissolved solids	N.R.	9,190	2,190	–	11,254	2,075

* No age of fill specified for Sources 1 through 3, Source 4 is initial, 5 is 3 year old, and 6 is from a 15 year old fill.

+ Data from "Development of Construction and use criteria for Sanitary Landfills," Report on USPHS grant DO1-U1-0046 Los Angeles County, California, 1968.

‡ EMRICH, G.H. and LANDON, R.A. "Generation of Leachate from landfills and its subsurface movement," Proceedings of the Northeastern Regional Antipollution Conference University of Rhode Island, Kingston, July 1969.

Leachate character reflects a host of interrelated physical, chemical, and biologic factors that control decomposition; refuse composition, compaction, average particle size, free oxygen availability, operation method; ambient temperature and hydrologic conditions. All of these variables are important. Landfills are often conceived of as homogenous units, but internally the degradation environment alters, aerobic breakdown continues in upper layers weeks after placement, but anaerobiosis is quickly reached at depth. Water movement may be channelled along refuse cell boundaries not unlike englacial streams, making it difficult to define a portion of a landfill at any given time.[16]

Recent research has concentrated on laboratory lysimeter studies.[17] Refuse is packed in artificial cells, doused with water, and leachate collected; character and concentration variations are correlated to environmental parameters. Such work may be useful if results can be applied to planning landfills, enabling predictability of leachate problems at specific sites. Some workers, however, feel this approach to be conceptually erroneous.[18] In seeking to bring lysimeters to field capacity quickly by regularly adding water, a picture of rapid decomposition with concentrated leachate over a short time period is given. In reality, leachate exists as a potential contaminant for many years, even after landfilling operations have finished and final cover has been placed. Schlinker for example, has reported German experience where wells were contaminated by an old covered refuse dump, 3,500' upstream. At a flow velocity of 2.5' per day, it had taken 6 years for leachate to migrate that distance and contaminate the water supply.[19]

For leachate to be produced, water must enter the landfill by at least one of three mechanisms; percolating rainfall, runoff channelled onto the site, or through groundwater movement where part of the fill lies below the water table. Once field capacity is attained leachate percolates vertically downward to the top of the zone of saturation if the medium is

TABLE 4,7

SUMMARY OF PHYSICAL, GEOCHEMICAL AND
BIOCHEMICAL PROCESSES THAT CONTRIBUTE
TO THE RENOVATION OF SOLID AND LIQUID MATERIALS

PROCESSES		
Physical	Geochemical	Biochemical
Diffusion	Complex Ion Pair Formation	Solute Uptake in Bio-synthesis
Filtration	Acid-Base Reactions	Solubilization of Cell-ulose, etc.
Dispersion		
Dilution	Inorganic Redox Reactions	Mineralization of Organic Redox Reactions
Adsorption-Desorption	Ion Exchange	
Gas Transfer	Precipitation Solution	

Source: PARIZEK, R.R. "Site Selection Criteria for Waste Water
Disposal--Soils and Hydrogeologic Considerations" in
Proceedings--Recycling Treated Municipal Waste-Water and
Sludge Through Forest and Cropland. Edited by SOPPER, W.E.
and KARDOS, L.T., Pennsylvania State University, 1973, p. 98.

homogenously isotropic, then following groundwater in the direction of
potential gradient.[20]   As leachate migrates through the ground it is
attenuated by physical, geochemical and biochemical processes. This may
occur very effectively, some environments have an almost miraculous renovative capacity, as demonstrated by the few reported contamination cases.
Forecasting attenuation rates involves consideration of local geomorphology,
leachate character and the interrelated processes outlined in Table 4,7.
The complexity of this task has so far hindered efforts to produce a predictive
mathematical model. Some research workers estimate that this will be
developed in the next few years.[21]   Currently in use are relatively simple

empirical models drawing upon dispersion, dilution and ionic exchange
as the prominent renovative processes.

Dispersion is a physical process varying with the nature and
regularity of regolithic pores through which liquid passes.[22] Contaminants
move as a slug within native groundwater,[23] dispersion operating along
the advancing front,[24] but there is little mixing for flow is laminar not
turbulent. Thus dispersion proficiency reflects the total area over which
contaminated water is in contact with native water. This mechanism and
resulting dilution is more efficient if the contact area is large. In media
where clay lenses or boulders are entrapped within a fairly uniform matrix,
mixing is helped as the slug is 'broken up' in navigating such obstacles,
simultaneously the contact face with native water is enlarged. This process
may be more influential in vertically downward infiltration through aerated
zones than in lateral movement within the zone of saturation. When liquid
moves through fractured or jointed bedrock it is spatially constrained and
there is little dispersion or dilution. Carbon-iferous limestone presents the
case, water moving along joints and solution lines, so that contaminants
may migrate miles in concentrated form.[25]

Ion exchange is an geochemical process; earth materials remove
and absorb oppositely charged ions from adjacent solutions.[26] Surficial
material appears to act somewhat like a wet sponge, absorbing and
filtering liquid in transit, yet unlike the purely physical process of filtration that removes suspended solids. Exchange capacity increases directly
with particle surface area, clays have the greatest values, clean gravels
the least.[27] Fractured or faulted bedrock, by concentrating leachate
during transit, renders ion exchange ineffective. Ion exchange generally
occurs slowly, the process taking place most readily in clays with high
relative permeability that hold liquid in contact with particles long
enough for the reaction to occur. This reinforcement is absent in pervious
material where leachate travels too quickly, and in impermeable clays
where movement is essentially zero. Silty or sandy clay with permeability

about $10^{-5/-6}$ cms./sec. optimizes this process, allowing effective chemical filtration while fostering dispersion and dilution.[28]

THE PHYSICAL BASIS OF SITE SELECTION MODELS

Environment and public health are usually protected by landfill engineering techniques not site selection and planning. However, this latter newer approach is presently being advocated by a minority of decision makers.[29] Engineering palliatives; impermeable covers, liners, drains and runoff diversion channels are valuable, being necessary where dumps become 'sanitized' or where leachate is collected for treatment before release. The author advocates a wider research base to this potentially disruptive land use, detailed physical investigations of urban/rural fringe land being desirable. In this way sites can be located where renovative processes are optimized, protecting both environment and public through nature's negative feedback (self repair) mechanisms. Such an analytical procedure may in fact guide engineering needs by illustrating renovative limitations. This aim is intuitively sensible, and will no doubt be more widely realized when additional research is undertaken into physical constraints. Basically the stumbling block is methodological, how best to model the factors involved in defining waste assimilative environments, particularly within the dynamic framework of the hydrologic cycle.

A number of workers are making progress in this field. Cartwright and Sherman analysed drift material from well records and surficial sediment and bedrock geology maps in Illinois to evaluate the State's waste receptive ability.[30] Using five criteria; type and thickness of unconsolidated material, bedrock nature, local and potential water sources, and site topography, they compiled a probability map relating geologic conditions to sanitary landfill feasibility. Three classes were recognized, generally unfavourable (thin cover and shallow acquifers), locally favourable (acquifer conditions variable), generally favourable (shallow acquifers rare).

Indentification of similar land units was advocated by Gartner, but his
pollution potential map expressing ground and surface water relationships
with regolith properties is biased towards minimizing technical and construction problems.[31] Fischer and Woodford expressed the aim of such research succinctly, excellent sites identified through geotechnical and
ecological knowledge provide "natural protection against pollution, "
poor sites need "engineered safeguards or remedial operations to upgrade
protection."[32]

Morekas reported an advanced empirical system of site rankings
designed to assess the U.S.A.'s capacity for long term storage and processing of hazardous waste; concentrated toxicants, radioactive and
biologic material.[33] A working philosophy of designing with nature, the
wide range of physical criteria analysed and factor weighting methodology
used, could all be transferred to land disposal of less hazardous municipal
solid waste. The 3,050 counties of the coterminous U.S. form the analyzed
unit, but to include risk and transportation costs, 36 regions based on equal
distance from common waste sources were built into the model. As a result, cross regional comparisons are invalid. Four general criteria for
site classification were reported by Morekas.

1. Earth Sciences and Climatology (involving physical and chemical
   considerations),
2. Transportation,
3. Ecology,
4. Human Environment and Resource Utilization.

These criteria were evaluated by ten specialists. Each of whom
scored every county on a scale of 0-5 with respect to the criterion with
which he had the greatest familiarity. A score or 5 being assigned where
that particular criterion was the most suited for hazardous waste disposal,
1 where conditions were least satisfactory, and 0 denoting a totally
unsuitable site with a potential for creating crisis. The team then individ-

ually weighted the four criteria in importance, final ratings used in analysis are shown in Table 5, 7.

TABLE 5, 7

HAZARDOUS WASTES WEIGHTING FACTORS

		Geology	7
		Hydrology	8
		Climatology	10
		Soils	6
Earth Sciences	31		31
Transportation	28		
Human Environment & Resource Use	23		
Ecology	18		
Total Criteria-	100		

The sum of the weighting factors (0-100) multiplied by the ranking value (0-5) gave every county a quantitative rating expressing its suitability for storing and processing hazardous wastes. Comparison may be made to a perfect score of 500 and scores of other counties in the same region.

This methodology is interesting because it contains several innovations. Ranking values are assigned by professionals working in their own field, hopefully similarly qualified and informed specialists would repeat the values given. The weighting factor is subjective, reflecting individual and group views as to the relative importance of each criteria. Other groups would, no doubt, weigh criteria diff-erently, however, this variation should diminish with increasing numbers and diversity.[34]

This weighting concept appears applicable to sanitary landfill
site selection. Here, transportation is of less concern as are human
environment and resource use, particularly resource foregone due to
disposal site location.

In 1973 Pavoni et. al. published on advanced landfill site
ranking system, a methodology to "evaluate the potential danger of
depositing any waste or material in a particular site."[35] Evaluation
is founded upon the following three site and waste characteristics:

1. potential for precipitation and runoff to infiltrate the deposited
    wastes,

2. potential for the waste material to be transported through fluid
    transmission away from its deposit location through underlying
    soils to groundwater systems and

3. other mechanisms for the removal of hazardous materials from the
    site and their transport to other areas.

In this study ten variables were quantified; infiltration potential (Ip), bottom leakage (Lp), filter capacity (Fc), and adsorptive
capacity (Ac) describing the site's soil system; organic content (Oc),
buffering capacity (Bc), travel distance (Td), and groundwater velocity
(Gv) delineating site groundwater characteristics; wind direction (Wd)
and population factor (Pf) depicting air parameters.

The weighting concept is introduced by scoring these variables
on four separate numerical scales according to their assumed relative
importance. Parameters immediately affecting waste transmission;
infiltration potential, bottom leakage and groundwater velocity,
were assigned "first order priority;" those affecting waste transmission
after contact with water; filtering capacity and adsorptive capacity were
given second priority. Third order priority was given to organic content
and buffering capacity; representing conditions of receiving groundwater. Parameters outside the immediate disposal site, travel distance,

wind direction and population were given the lowest, fourth order priority. A maximum value of 20 units was arbitrarily assigned to first order variables, for second, third and fourth orders, maximum values were 15, 10, and 5 respectively. Parameter values are calculated from fairly simple formulae, denominated to bring values within the designated scoring ranges 0-20, 0-15, 0-10 or 0-5. Any landfill may be characterized by values from approximately 0-110, lower numbers represent better waste reception environments.

This model excludes engineering bias and allows the rating of natural characteristics giving it applicability in climatic and geomorphic extremes. The ten variables considered cover most of the leachate formation and attenuation processes, although some only indirectly. The scheme necessitates ten calculations for each data point if analysis is to be spatially oriented, and also raises the perennial spectre of data availability. When a short list of three or four disposal sites is examined, first hand data collection is essential, but in categorizing many square miles, readily available pre-existing data, suitable for numerical conversion, is needed. This model becomes extremely unwieldy if analysis of hundreds or thousands of data points is required. It was designed to rate proposed sites, not to assess large areas of the rural/urban fringe to find suitable disposal locations. However, if data can be obtained and computer analysis is used to manipulate it, this comprehensive model is probably the best yet available.

In 1963, before the growth of ecological awareness, Le Grand put forward a semi-quantitative model for evaluating waste disposal sites. Perceptively he realized the topic cut across academic boundaries, traditional specialists; geologists, biologists and engineers were poorly equipped to synthesize data and appraise sites. To overcome this difficulty and help health officials charged with enforcing rigid waste disposal guidelines he presented a system to "evaluate contamination potential of areas where wastes are released in loose granular earth, at or near the surface."[36] This model was designed to give decision makers a preliminary site evaluation where geologic and hydro-logic data was scarce. Significantly the LeGrand system was not intended to replace on-site fieldwork.

Sites are characterized in terms of the probable effects that five environmental factors have on released wastes. It's applicability is wide, designed for contaminants that "attenuate or decrease in potency in time or by oxidation, chemical or physical sorption and dilution through dispersion."[37] This definition includes sewage, detergents, viruses, and radioactive matter. Although not designed for evaluating environmental impact of solid waste leachate it seems likely that it can be sucessfully used for this purpose if one or two extra parameters, notably the amount of water input are included. LeGrand noted it should not be used where the "critical (emphasis added) consideration is the movement of chemical wastes that attenuate slowly."[38] Some leachate components, for example chloride and nitrate are covered by this warning, but landfill leachate is one of the most complex of liquid wastes, and may contain a wide variety of other substances.[39] Any leachate renovation model will have individual components behaving against general trends, a fact that should not jeopardize attempts to formulate and test comprehensive models. Since the aim of this study is to areally categorize the ability of the Saanich Peninsula to renovate leachate, to be followed by site selection through fieldwork, this methodological problem appears relatively insignificant.

Three disposal environments were recognized by LeGrand:

1. unconsolidated granular material extending 100' or more below ground (one media site),

2. unconsolidated granular material at ground surface underlain at shallow depths by dense rocks with linear openings (two media site) and

TABLE 6,7

GROUNDWATER CONTAMINATION POTENTIAL RATING CHART FOR USE IN LOOSE GRANULAR MATERIALS (ONE MEDIUM SITES)

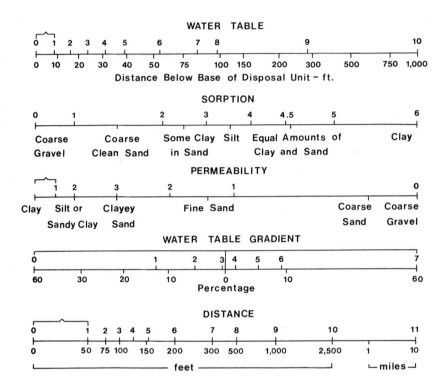

On all scales the point values are indicated by the upper scale; the brackets indicate unacceptable ranges for any factor, except the two brackets on the gradient scale, one labeled AD, which is for an adverse direction of flow (toward point of water use), and one FD, which is for a favourable direction of flow.

Source: H.E. LeGrand, "System for Evaluation of Contamination Potential of Some Waste Disposal Sites" *Journal of the American Water Works Association* Vol. 56 (August 1964) p. 964.

TABLE 7,7

GROUNDWATER CONTAMINATION POTENTIAL RATING CHART
(TWO MEDIA SITES)

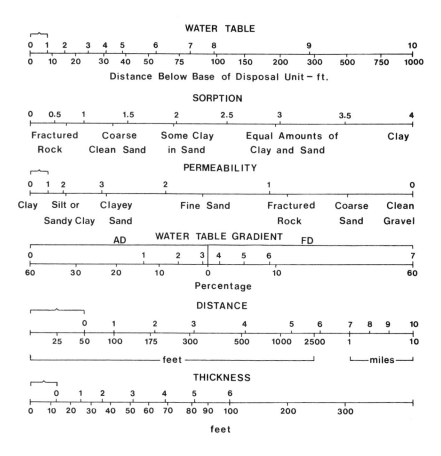

On all scales the point values are indicated by the upper scale; the brackets indicate unacceptable ranges for any factor, except the two brackets on the gradient scale, one labeled AD, which is for an adverse direction of flow (toward point of water use), and one FD, which is for a favourable direction of flow.

Source: H.E. LeGrand, "System for Evaluation of Contamination Potential of Some Waste Disposal Sites" *Journal of the American Water Works Association* Vol. 56 (August 1964) p. 966.

3. dense rocks at ground surface, with fluid movement along joints or solution channels.

The five factors considered in his model were water table depth and gradient, surficial material permeability and sorption and distance to the nearest point of water use. Each factor is connected to a numerically rated scale, at any data point natural conditions are converted by reading from scales provided (Table 6,7 and 7,7). Values are then summed, giving a groundwater contamination potential figure to be interpreted as shown in Table 8, 7.

The weighting concept is introduced so that the scales do not increase in simple arithmetic progression but vary with each factor (Table 6,7 and 7,7). Maximum possible points for one medium (unconsolidated sediment) and two media (includes bedrock) sites is given below, note the overriding importance of water table depth and surficial material thickness in the later. This mirrors the great reliance placed on the zone of aeration for most of the attenuation processes listed in Table 4,7. The weights attached to the five environmental variables are listed in Table 9,7.

TABLE 8,7

CONTAMINATION POTENTIAL

0 - 4	Pollution imminent
4 - 8	Pollution probable or possible
8 - 12	Pollution possible but unlikely
12 - 25	Pollution very improbable
25 - 35	Pollution impossible

TABLE 9,7
WEIGHTING OF ENVIRONMENTAL FACTORS
ONE MEDIUM SITE

Factor	Maximum Points	Per cent of Total
Water table depth	10	27
Sorption	6	16
Permeability	3	8
Water table gradient	7	19
Distance to nearest point of water use	11	30
	37	100

TWO MEDIA SITE

Factor	Maximum Points	Per cent of Total
Water table depth	10	25.0
Sorption	4	10.0
Permeability	3	7.5
Water table gradient	7	17.5
Distance to nearest point of water use	10	25.0
Unconsolidated sediment thickness	6	15.0
	40	100.0

In the LeGrand model contamination potential decreases with distance between waste source and the nearest point of water use since the environment is given time and space to renovate effluent. Such assumptions are valid because:

1. Dilution increases with distance.

2. Sorption (ion exchange) tends to be more complete with distance.

3. Travel time increases with distance, thus decay or degradation is more complete.

4. Water table gradient and consequently flow velocity tend to decrease with distance.[40]

LeGrand's scheme is empirical but quantitative studies were used in scale preparation which were then refined by repeated trial, adjustment and field checking. The accuracy of predictions based on this model are claimed to be good for one medium sites and fair to good for two media.[41] The weighting system has been arbitrarily constructed,[42] but more modern models have also been unable to avoid this subjective element. The LeGrand system is easily handled, since the necessary data is likely to be available from well logs; conversion is also extremely simple as scales are used. It can be adapted for use to show variations over space since distance affects the efficiency of all renovative processes. However, with modification, dynamic variables, rainfall, runoff and evapotranspiration can also be incorporated. For these reasons the author concluded that the LeGrand's model was the best available for use in an attempt to evaluate part of the British Columbian Capital Regional District's ability to assimilate municipal solid waste.[43]

REFUSE DISPOSAL IN GREATER VICTORIA

In 1971 Metropolitan Victoria had 200,000 inhabitants, over 95 per cent of the Capital Regional District's total population. By 2001 it is expected to exceed 350,000.[44] Per capita refuse production, estimated to be 2.63 lbs a day, is extremely low compared to other North American centres.[45] This fact partially reflects community attitudes; refuse, especially litter, conflicts with the parochial but commendable pride residents feel for their "garden city". Much garbage is composted and used in gardens, many of which are tended by retired couples attracted to Victoria by its mild climate. Many homes are single dwelling units with gardens, enabling on site composting or incineration of at least some refuse. The accelerating trend to condominiums will prevent these processes, no doubt necessitating greater collection, organization and central disposal of domestic solid waste. There is very little industry in Greater Victoria. The provincial government, a major employer, internally recycles its own waste paper. In outlying districts fewer convenience products are consumed and greater potential exists for personal disposal, accounting for lower per capita generation rates.[46] However, over 250 tons of refuse is collected for disposal in Greater Victoria daily, by 2001 this may reach 600 tons[47] (Figures 1,7 and 2,7).

All refuse collected by municipalities and private firms is eventually landfilled at the Hartland Road dump, eleven miles northwest of downtown Victoria. This site has been used since 1959 but was converted to a sanitary landfill operation in 1968 at the insistence of the Greater Victoria Metropolitan Board of Health. Simultaneously, smaller dumps to the west of the city in the Millstream and Craigflower Creek areas were closed, their refuse being diverted to Hartland. About 117,000 tons of refuse is accepted annually, together with 50,000 cubic yards of cover material, a varying percentage of which is hog fuel, (bark chips) from a local sawmill and plywood plant.[48]

The landfill is located in an upland north-south aligned valley (Figure 3,7). Landfill operations are gradually filling it to a depth of between 20 and 30 feet, a ramp variation being used. The valley is incised into acidic igneous rocks, mostly diorite and gneiss covered by thin, patchy deposits of unconsolidated glacial and fluvio-glacial material. Bedrock is only slightly faulted, groundwater movement through fissures is low, as is percolation; porosity and permeability are slight. Surrounding slopes are open woodland with many bare rock outcrops.[49] As a result of its topography and bedrock geology, the valley without the landfill would have a high runoff coefficient. Annual rainfall is quite high, about 44 inches with winter maxima of 9 inches per month in December and January.[50]

The landfill surface area was 26.5 acres in June 1972 and about 30 acres in the summer of 1975. The size of the original drainage basin, now

FIGURE 1,7
Input: refuse transfer station, Victoria.

FIGURE 2,7
Output: refuse transfer station, Victoria.

FIGURE 3,7
The Hartland Road sanitary landfill.

containing the Hartland dump was 200 acres but this was reduced to 30 acres when runoff diversion channels were built to separate it from Mt. Work to the west.[51] In March 1975 these were found to be in poor repair and, as a result, a high percentage of runoff was escaping into the fill. If it is assumed that the landfill has a surface area of 26.5 acres and that the effective rainfall (total rainfall minus evaporation) is 2.75 feet annually, then the landfill is receiving a maximum water input of 621.8 acre feet per year and a minimum of 155.3 acre feet annually. The actual volume entering the landfill will depend upon the effective size of the drainage basin, which in turn is reflecting the efficiency of the artificial diversion channels. Little leachate escapes to groundwater, nearly all emerges at the fill's northern end and is released into Heal Creek, flowing 2 3/4 miles to the sea at Tod inlet, through Durrance and Tod Creeks (Figure 4,7).

In 1970 and 1971 representatives from Butchart Gardens complained that water dammed from Tod Creek and used for irrigation purposes was contaminated with a putrid slimy growth. Examination of the yellow – to – brown slimes which developed on solid surfaces showed they were mainly iron oxide, probably resulting from the action of iron fixing bacteria that utilize soluble iron in water as parts of their metabolism. This prompted investigation of the landfill leachage, released two miles upstream, by the provincial Pollution Control Board and by B.H. Levelton and Associates Limited, environmental consultants under contract to Victoria Disposal Limited, the landfill owner and operator.[51]

Studies determined the leachate was warm, black, with a musty anaerobic smell, and like that examined elsewhere contained many contaminants.[95] Volume was seasonally controlled; peaks of 200 gallons per minute being found in wet winters and lows of 0 – 10 gallons per minute during dry summers. Dissolved iron (ferrous), insoluble iron (ferric hydroxide), and B.O.D. were major contaminants. To remedy the downstream problems caused by this leachate two holding ponds were built in 1972, the first with a 40,000 imperial gallon capacity was aerated by a portable com-

FIGURE 4,7
Leachate entering Heal Creek.

pressor to oxidize and precipitate dissolved iron to its ferric form. The second, capable of holding 170,000 imperial gallons removed B.O.D. and continued settling of ferric hydroxide and suspended solids.[52] (Figure 5,7). The effectiveness of this treatment system depends on the throughflow time, during high runoff, that is above 150 gallons per minute, there is insufficient time for proper B.O.D. removal, fortunately such peaks correspond to natural streamflow peaks and dilution by surface water is very effective. Since this crude treatment 'system' was introduced there have been no downstream complaints or recurrences of algae 'blooms' at the Butchart Gardens reservoir.

This landfill is located in an active geomorphic environment, leachate volume, despite preventative measures, is high due to heavy rainfall and runoff. There is no threat to groundwater as bedrock, which is at or very near the surface, is relatively impermeable, confining leachate to the surface. Here it is released into streams after crude treatment but great reliance is placed on mixing and dilution to render it potable to downstream users. This process is a success since users are presently far enough downstream, but suburbia has been encroaching into the area for many years and problems are likely to increase as development continues.[53] Most of the renovative processes listed in Table 4,7 do not occur in this case. As people settle closer to the Hartland landfill this fact will become increasingly obvious and sophisticated and expensive treatment will be needed to supplement natural dilution. However, with only five years landfill space available, a new site will of necessity be in operation before this occurs.

The Capital Regional District which assumed responsibility for refuse disposal but not collection in 1973 has received recommendations for new landfill sites but these remain confidential. Any site will have to serve all Greater Victoria, accepting a minimum of 250 tons of refuse per day. Leachate contamination will continue to be a problem for rain-

FIGURE 5,7
Leachate settling ponds.

fall exceeds actual and potential evapotranspiration throughout the
Capital Regional District. Its significance may be minimized if site
selection includes consideration of both physical and socio-economic
variables (Figure 6,7).

SAANICH PENINSULA LANDFORMS

The topography of the Saanich Peninsula reflects the action of a
variety of geomorphological processes; glacial erosion and deposition and
eustatic and isostatic adjustment have been important, partially accounting for the region's distinctive relief. Unconsolidated sediments are
currently being eroded by fluvial and marine processes, especially along
lines of least resistence.

The bedrock geology of the area was initially investigated by
G.M. Dawson in 1876,[54] his study was followed by the definitive work
of Charles Clapp in the early twentieth century. A wide variety of bedrock types is found in the Capital Regional District, ranging from fine
grained sandstone, to metamorphic, and intrusive volcanic. Saanich
granodiorite, the principle intrusive type occupies most of the peninsula.
This is well jointed and fractured, yielding up to 10 gallons per minute in
wells that penetrate fracture zones.[55] West of a line, connecting Tod
Inlet, Prospect Lake and Esquimalt Inlet, bedrock is at or very close to
the surface, elsewhere it is masked by surficial deposits. Bedrock remnants
standing above the surrounding erosion surfaces are visually prominent,
Mt. Tolmie (408 feet), and Douglas (739 feet), show rounded and smoothed
profiles, evidence of debris ladened ice erosion during the Fraser glaciation.

Surficial materials are all thought to be Pleistocene in age.[56]
Glacial deposits consist of till, clay, silty clay, sand and gravel,
deposited in continental and marine environments directly by ice or by its
meltwaters.[57] Interglacial deposits are mainly fluvial, consisting of gravel,
sand, silt and peat. Sequence and thickness of deposits reflect the area's

FIGURE 6,7
The Hartland Road sanitary landfill in operation.

geomorphic history. [58] The maximum thickness of surficial sediments recorded in well logs, is about 400 feet. In rare instances, 200 feet of sediments are exposed in cliff sections although normally 10 feet – 60 feet is expected. Most wells receive water from unconsolidated deposits and do not penetrate bedrock; the yield varies greatly depending on acquifer quality. The most productive acquifer is the Quadra sands from which over 150 gallons per minute may be obtained. Deposits are spatially patchy, complicating attempts to model the depositional character of the region, an indirect aim of this study.

APPLICATION OF LEGRAND'S SYSTEM TO THE SAANICH PENINSULA

The most comprehensive data source available is that assembled by the Greater Victoria Environmental Geology Survey in 1972. This programme, sponsored by the Geological Survey of Canada, collated 3,100 well and engineering logs, covering the Saanich municipalities and Gulf Islands Electoral District. Information came from three sources; Groundwater Division, B.C. Water Resource Service, Geotechnic and Materials Testing Branch, B.C. Department of Highways, and Thurber Associates Limited, Geotechnic Engineers, Victoria. This information is held on open file at the Capital Regional District, Building Inspection Department offices at Colwood, a duplicate set is held by the Geological Survey in Ottawa.

Information detail varies with the drilling contractor originally recording the log. Twenty-four contractors supplied logs, but not on a uniform basis. This raised a categorization problem; silty clay to one was perhaps clayey silt to another! Beside details of surficial sediment en- countered in drilling, bedrock geology, groundwater depth, and yield, occasional mention was also made of groundwater quality and soil properties. The accuracy of this information is difficult to ascertain and no doubt varies markedly, but is comprehensive, recent, and the only available source suitable for the study in hand. Aerial photographs were used to check information where possible.

This data was converted to the LeGrand numbers as previously discussed, (Table 6, 7 and 7, 7) so that the spatial variations in the suitability of part of the Capital Regional District for waste disposal could be assessed. This involved a series of steps:

1. Data point locations were plotted on a 1:25,000 base map from military grid reference identification used by the survey. This scale was chosen because it was close to that used in the 1973 Capital Regional District Land Resources Inventory Programme. Of 3,100 records in the survey, 1,104 were within this study's boundary, the Saanich Peninsula north of urban Victoria. During transfer, some data points were eliminated; those where information was so sketchy as to be meaningless, and those where similar values were clustered preventing separate identification at the base map scale. As a result, 986 well log records were utilized.

2. Using the two media rating chart shown in Table 7, 7 each log's information was converted to LeGrand numbers, a separate data set being kept for sorption, permeability, surficial material thickness, and water table depth. The latter environmental variable presented a problem, being unrecorded on engineering records. Well logs were poorly compatible; seepage depth, and settling level after 12 or 24 hours were all intermittently recorded. Since 1969, water table level in three of the four automatic observation wells monitored in the peninsula by the Groundwater Division has dropped dramatically reflecting excessive extraction rates.[59] Extension of the Greater Victoria Water Board surface water supply system to North and Central Saanich would replace many wells, allowing the groundwater to recover and invalidating analysis based on post-1969 records. Since recent overpumping appears to have artificially lowered the water table over much of the Saanich Peninsula, the pre-1969 position of the water table was estimated using information presented by E.B. Wiken from aerial photographs and land unit analysis. This was supplemented by data from pre-1969 well logs.[60]

    The distance to nearest point of water use is included in the LeGrand system but was excluded from data point numeration since it is not a physical attribute.

3. Water table gradient, seemingly the most arbitrarily scaled of the six factors, could not realistically be included, being defined to the nearest point of water use, already excluded. Yet it was vital to retain integrity of the contamination score guide of 0-35 with its qualitative

271

ratings. Two methods of overcoming this difficulty appear feasible; it is possible to assume each data location is in an adverse flow direction of over 20 per cent, thus giving every point a zero score for this variable, maximizing environmental and public health safety but probably reducing the size and number of areas that could safely assimilate waste. Alternatively, every data location could be given three points, representing parallel flow direction; neither adverse nor favourable. Both options are presented in the map interpretation tables (Tables 10,7 and 11,7).

4. Symap (version five) a computer mapping programme, developed at the Laboratory for Computer Graphics and Spatial Analyses at Harvard University and subsequently modified by E.D. Dudnik, Department of Architecture, University of Illinois was used to portray the data.[61] This programme produces maps which graphically depict spatially disposed qualitative and quantitative data accurately, and in a manner allowing comparative analysis. A valuable tool, it is now being increasingly used by resource managers.[62]

Tables 10,7 and 11,7 to be used for interpretation of the maps presented as Figures 7,7 and 8,7 allow weight to be given to the following numerated factors. These are water table depth, suficial material thickness, sorption and permeability upon which computer map shading interpolation is based; water table gradient at values of 0 and 3 points; contaminant migration at distances of 0-50', 100', 500', 1200' and one mile; and a pollution rating scale (0-35 points) shown in Table 8,7. The pollution rating scale has 4,8, 12, and 25 as its threshold numbers. These separate five contamination categories portrayed in Tables 10,7 and 11,7 with neutral (G = 3) and adverse (G = 0) water table gradients. Attenuation increases with the distance travelled by contaminants, renovation therefore increases with distance. This is clearly seen from the step-like nature of the shaded contamination categories. Once these principles have been grasped Figures 7,7 and 8,7 can be interpreted. If a well is 100 feet from a sanitary landfill, located in map shade 2, it can be seen from Table 10,7 that contamination is probable or possible. Similarly if a well is 500' from a landfill, sited in map shade 3, contamination is possible but un-likely. Table 11,7 is read similarly, although base values are higher since a

272

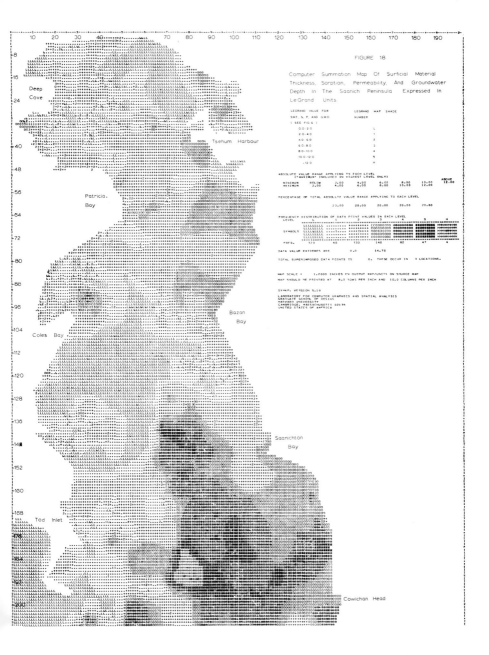

FIGURE 7,7   Leachate renovation capacity, Saanich Peninsula.

273

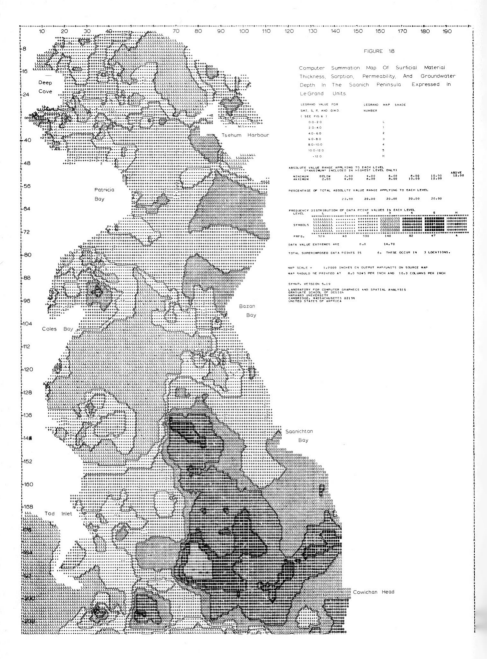

FIGURE 8,7   Contoured renovation capacity, Saanich Peninsula.

neutral water table gradient (G = 3) is incorporated. It should be stressed that LeGrand at no stage suggested this method could replace fieldwork in selecting disposal sites. The main advantage of the spatial analysis reported here is that it helps to identify and so eliminate grossly unsuitable regions, enabling fieldwork to be concentrated where physical conditions are favourable.

This model can be improved if the impact of climatic factors on volume or potential volume of leachate can be included. Leachate contamination intuitively becomes increasingly likely the more annual precipitation exceeds evapotranspiration. Seasonal variations in climate are important since heavy winter rainfall or spring snowmelt may reduce the ability of the environment to assimilate leachate, resulting in surface water contamination.

The author considers that mean annual potential runoff (the sum of monthly precipitation minus monthly evapotranspiration) is perhaps the best climatic parameter to model. This represents water that is actually available to enter the landfill, although it would be increased if the land-fill was sited in a drainage basin which supplied it with additional runoff and decreased by surface contouring to diminish percolating rainfall. Pavoni would presumably have categorized this variable as "first order" since it immediately affects waste transmission. Therefore it should be given greater weight than sorption, permeability, and sediment thickness in categorizing environmental ability to renovate leachate, being best represented by 7 or 8 points, on a forty point scale, (18-21 per cent) see Table 9,7. If this factor is added to the two media scale shown in Table 7, 7 a maximum possible score of 47 or 48 points is obtained. If the scales are altered in this way the contamination score guide will need re-structuring, alternatively existing factor ratings should be reduced, enabling potential runoff to be included within present score categories. Both methodological possibilities are being examined.[63]

If precipitation is taken into consideration a potential runoff
scale of 0-8 points on a semi-logarithmic base, similar to that used to rate
the significance of water table depth should be developed (Tables 6,7 and
7,7). Maximum points would be awarded in a zero runoff situation where
the sum of monthly precipitation is exceeded by the sum of monthly
evapotranspiration. Absolute measures in inches per year or acre feet per
year appear most meaningful.

FIGURE 9,7
RUN-OFF POTENTIAL SCALE

Mean Annual Potential Runoff (P.R.)

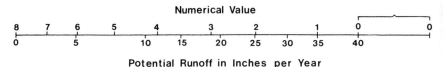

Potential Runoff in Inches per Year

The Greater Victoria area has a relatively simple orthographically
controlled preciptiation gradient that increases from east to west. Mean
annual potential runoff is about 20" in the extreme east and above 35"
in the western highlands where the Hartland Road landfill is sited,
(Figure 7,7). Using Figure 6,7 it can be seen that this represents a
difference of 1.5-2.0 points, on the potential runoff scale.

The computer map, a section of which is shown in Figure 7,7,
portrays the spatial distribution of the numerated factors, thus categorizing
the Saanich Peninsula's leachate renovation ability. Figure 8,7 shows
this information at a smaller scale, facilitating recognition of the renovation categories throughout the peninsula. From these maps a number of
conclusions may be drawn.

1. The Saanich Peninsula possesses a variety of geomorphic environments,
   some of which are unsuited to sanitary landfilling. However, much
   of the peninsula may be used for this purpose with minimal environmental degradation.

2. Of the 986 data points analysed, 34.5 percent fell within map
    shade L, 4.2 percent within map shade 1, 33.9 percent within
    2, 14.8 percent within 3, 8.2 percent within 4, 4.0 percent
    within 5 and 0.4 percent within H (Table 10,7 and 11,7).

3. Poor, that is low scoring terrain, is found where bedrock is at
    or close to the surface and where the water table is encountered at
    shallow depths. Generally the renovative environment improves
    eastwards. The western highlands and bedrock intrusions of Mt.
    Douglas, Mt. Newton, Cloake Hill, and Bear Hill are all shaded
    L. The highest scoring area is sited north of Bear Hill and is a
    wedge shape, three miles long north to south, two miles wide at
    its base and a quarter of a mile wide at its apex. Here surficial
    material is thickest and water table deepest. However, zones of
    sand and gravel within this area may make waste disposal proble-
    matic by over diffusing leachate. This adverse factor was out-
    weighed by high scores on the other variables, reflecting this
    methodology's weighting system.

4. Although not formally included in the map construction, climatic
    parameters reinforce its interpretation as potential leachate volume
    decreases eastwards with declining precipitation and potential run-
    off. The wedge shaped area identified as the best renovative en-
    vironment receives about 35" of precipitation per year with a mean
    annual potential runoff of 24" per year. This is relatively high in
    an absolute sense but low in the context of the whole peninsula.
    This relatively high precipitation also stresses the need for landfill
    engineering to maximize surface runoff and reduce infiltration to
    decomposing refuse (Figure 10,7).

5. This spatial analysis suggests that fieldwork should concentrate in
    areas shaded 3, 4, 5, and H, that is 27.5 percent of the peninsula
    examined. Areas L, 1 and 2 should be discarded as unsuitable for
    solid waste disposal. Fieldwork should be concentrated in the
    former areas placing greatest emphasis on water table gradient as
    this is the most poorly numerated factor in the present study.

6. Socio-economic considerations may eliminate this landuse from
    much of the area designated for detailed fieldwork. This does not
    detract from the rationale adopted by this study, that is, locating
    solid waste disposal operations where the ability of the environment
    to renovate leachate is greatest. However, it does indicate the
    need for a methodology that will integrate social and physical
    factors to form a comprehensive planning base toward this landuse.

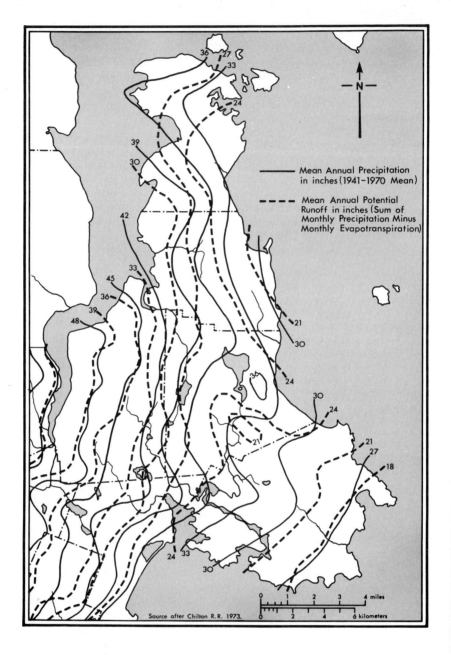

FIGURE 10,7   Precipitation and potential run-off.

278

TABLE 10,7
SYMAP INTERPRETATION (ADVERSE WATERTABLE GRADIENT)

MAP SHADE		LEGRAND NUMBER	CONTAMINATION POTENTIAL (L.G. NUMBERS) AT VARYING DISTANCES FROM THE WASTE SOURCE				
No.	Symbol	(S,P.Wt,T,&G=0)	0-50'	100'	500'	1200'	1 Mile
L	L	-2	-2	-3	-6	-7	-9
1	•	2-4	2-4	3-5	6-8	7-9	9-11
2	+	4-6	4-6	5-7	8-10	9-11	11-13
3	0	6-8	6-8	7-9	10-12	11-13	13-15
4	θ	8-10	8-10	9-11	12-14	13-15	15-17
5	■	10-12	10-12	11-13	14-16	15-17	17-19
H	H	+12	+12	+13	+16	+17	+19

0 - 4 = Pollution imminent
4 - 8 = Pollution probable or possible
8 - 12 = Pollution possible but unlikely
12 - 25 = Pollution improbable
25 - 35 = Pollution impossible

279

TABLE 11,7

SYMAP INTERPRETATION (NEUTRAL WATERTABLE GRADIENT)

MAP SHADE		LEGRAND NUMBER	CONTAMINATION POTENTIAL (L.G. NUMBERS) AT VARYING DISTANCES FROM THE WASTE SOURCES				
No.	Symbol	(S,P.Wt,T,& G=3)	0–50'	100'	500'	1200'	1 Mile
L	L	-5	-5	-6	-9	-10	-12
1	•	5–7	5–7	6–8	9–11	10–12	12–14
2	+	7–9	7–9	8–10	11–13	12–14	14–16
3	O	9–11	9–11	10–12	13–15	14–16	16–18
4	θ	11–13	11–13	12–14	15–17	16–18	18–20
5	▪	13–15	13–15	14–16	17–19	18–20	20–22
H	H	+15	+15	+16	+19	+20	+22

0 - 4 = Pollution imminent
4 - 8 = Pollution probable or possible
8 - 12 = Pollution possible but unlikely
12 - 25 = Pollution improbable
25 - 35 = Pollution impossible

The predictability of environmental contamination can only be as accurate as the methodology used. Great improvements can be in- visaged in the model discussed in this paper. The most urgent need is to clearly define the relationships between factors and examine their multiple effects as leachate renovators. LeGrand portrayed factors as being directly or inversely interrelated.[64] The strength of these re- lationships require quantification through theoretical and experimental examination. If this is done, improved models will become available, increasing the degree of confidence with which contamination can be predicted.

REFERENCES

1. ZANONI, A.E. "Potential for Groundwater Pollution from the Land Disposal of Solid Wastes," Critical Reviews in Environmental Control, Chemical Rubber Company, 3, No. 3, (1973), p. 227.

2. Ibid., p. 228.

3. THURLOW et. al. A Preliminary Overview of the Solid Waste Problem in Canada. Victoria, Thurlow and Associates Limited. Report submitted to The Resources Research Centre, Political Research and Coodination Branch, Department of Fisheries and Forestry, Ottawa, April 1971, p. 21.

4. DARNAY, A. and FRANKLIN, W.E. The Role of Packaging in Solid Waste Management 1966-1976. U.S. Dept. of H.E.W., Public Health Service, Bureau of Solid Waste Management, Publication (SW-5c), 1969, pp. 12-13.

5. Editorial, "Turning Junk and Trash into a Resource," Business Week, No, 2145, October 10, 1970, p. 67.

6. Waste Management and Control, A Report to the Federal Council for Science and Technology by the Committee on Pollution National Academy of Sciences – National Research Council Publication No. 1400 (Washington D.C. 1966) p.35.

7. See DUNN, L.O., "Town and Village Teamwork," The American City, 76, No.5, May 1961, p. 108 for an account of open dump conversion to sanitary landfill.

8. "Sanitary Landfill, "American Society of Civil Engineers Manual, No. 39, (1959).

9. GRAINGE, J.W., EDWARDS, R., HEUCHERT, K.R., and SHAW, J.W. Management of Wastes from Arctic and Sub-Arctic Work Camps. Report No. 73-19 for the Environmental – Social Committee Northern Pipelines, Task Force on Northern Oil Development, September 1973, pp. 26 - 27.

10. A Handbook for Sanitary Landfills in Florida, Report Supplement to State of Florida Solid Waste Management Plan,

Department of Health and Rehabilitative Services,
Division of Health, Jacksonville, Florida, 1971.

11. SORG, T.J. and HICKMAN, H.L. Jr. Sanitary Landfill Facts.
Report (SW-4TS), U.S. Bureau of Solid Waste Management,
1970.

12. National Centre for Resource Recovery, Inc., Sanitary Landfill
A State-of-the-Art Study. Lexington Books, 1974, p.50.

13. HUGHES, G., TREMBLAY, J.J., ANGER, H. and D'CRUZ,J.
Pollution of Groundwater Due to Municipal Dumps. Inland
Water Branch, Dept. of Energy Mines and Resources,
Ottawa Tech., Bull, No. 42, (1971) p. 2.

14. ZANONI, A.E. op. cit., p. 234.

15. California State Water Pollution Control Board. Investigation
of Leaching of a Sanitary Landfill, Publication, No. 10,
1954, p. 13.

16. See the excellent discussion by SALVATO, J.A. , WILKIE, W.G.
and MEAD, B.E. "Sanitary Landfill-Leaching Prevention
and Control," Journal of the Water Pollution Control
Federation, 43, No. 10, (Oct. 1971) pp. 2084-2100.

17. QASIM, S.R. and BURCHIMAL, J.C. "Leaching from Simulated
Landfills," Journal of the Water Pollution Control Federation,
42, No. 3, (March 1970) pp. 371-379. See particularly
FUNGAROLI, A.A. Pollution of Subsurface Water by
Sanitary Landfill. Report (SW-12rg) U.S. Bureau of Solid
Waste Management (1971) pp. 13-34.

18. ROVERS, F.A. and FARQUHAR, G.J. Sanitary Landfill Study,
Volume 2, Effect of Season on Landfill Leachate and
Gas Production. University of Waterloo Research
Institute, 1972, p. 215.

19. SCHLINKER,K. Cited in California State Water Pollution Control
Board. Effects of Refuse Dumps on Groundwater Quality,
Publication No. 24, 1961, p. 90.

20. LeGRAND, H.E. " Management Aspects of Groundwater Contamination," Journal of the Water Pollution Control
Federation, 36, No. 9, (Sept. 1964) p. 1139.

21.   ROVERS, F.A., NUNAN, J.P., FARQUHAR, G.J. "Landfill
      Contaminant Flux-Surface and Subsurface Behaviour."
      Proceedings of the 21st Ontario Industrial Waste
      Conference, Toronto, Ontario. Ontario Ministry of
      the Environment, June 23 to 26, 1974, p. 107.

22.   SCHEIDEGGER A.E. distinguished between dispersion and
      diffusion. The former dependent upon pore system
      complexities, i.e. Material Nature, the later upon
      Intrinsic motion of Liquid Molecules. See "Statistical
      Hydrodynamics in Porous Media," J. of Applied Physics,
      25, (1954), pp. 994-1001 and 'General Theory of
      Dispersion in Porous Media', J. of Geophysical Research,
      66, No. 10, (Oct. 1961), pp. 3273-3278.

23.   LeGRAND, H.E. "Patterns of Contaminated Zones of Water in
      the Ground," Water Resources Research, 1, No. 1,
      (1965), p. 89.

24.   LeGRAND, H.E. "Monitoring of Changes in Quality of Ground-
      water," Groundwater, 6, No. 3, (1968), pp. 16-17.

25.   For a basic discussion see SCHBEIDER, W.J. "Hydrologic
      Implications of Solid-Waste Disposal," U.S. Geol.
      Survey Circular 601-F, 1970.

26.   ELLIS, B.G. "The Soil as a Chemical Filter," in SOPPER,
      W.E. and KARDOS, L.T. (eds.) Proceedings: Recycling
      Treated Municipal Wastewater and Sludge Through
      Forest and Cropland. Pennsylvania State University
      Press, 1973, p. 47.

27.   LeGRAND, H.E. "System for Evaluation of Contamination
      Potential of some Waste Disposal sites," Journal of the
      American Water Works Association, 56, No. 8, (Aug.
      1964), pp. 964-965.

28.   HUGHES, G.M. Hydrogeologic Considerations in The Siting and
      Design of Landfills. Illinois State Geological Survey,
      Environmental Geology Notes No. 51, April 1972, p.4.

29.   ROVERS, F.A., NUNAN, J.P. and FARQUHAR, G.J. op. cit.,
      p. 115.

30. CARTWRIGHT, K. and SHERMAN, F.B. Evaluating Sanitary Landfill Sites in Illinois. Illinois State Geological Survey, Environmental Geology Notes No. 27, August 1969.

31. GARTNER, J.F. "The Role of the Engineering Geologist in Pollution Control," Water and Pollution Control, 109, No. 11, (Nov. 1971), pp. 18-25.

32. FISCHER, J.A. and WOODFORD, D.L. "Environmental Considerations of Sanitary Landfill Sites," Public Works Part 1, 104, No. 6, (June 1973), pp. 93-97, Part 2, 104, No. 7, (July 1973), pp. 70-73.

33. MOREKAS, S. "Criteria for the Selection of Sites for Treatment and Disposal of Hazardous Wastes," Proceedings of the International Conference on Land for Waste Management. Ottawa, Canada, October 1973, pp. 308-316, Dept. of The Environment and National Research Council of Canada, Nov. 1974.

34. Ibid., p. 316.

35. PAVONI, J.L., HAGGERTY, D.J. and HEER, J.E. "Evaluation of Sanitary Landfill Sites," Public Works 104, No. 2, (Feb. 1973), pp. 55-57. More detailed information of the Ranking System is found in VAN NOSTRAND, Solid Waste Management, 1973. Also published as "Environmental Impact Evaluation Waste Disposal in Land," Water Resources Bulletin, 8, No. 6, (Dec. 1972), pp. 1091-1107.

36. LeGRAND, H.E. op. cit.

37. Ibid., p. 960.

38. Ibid., p. 960.

39. See discussions in Sanitary Landfilling Seminar: Proceedings, Environment Canada, Ecological Protection Branch, Solid Waste Management Report EPS 4-EP-72-2, December 1972.

40. LeGRAND, H.E. op.cit., p. 969.

41.   Ibid., p. 962.

42.   Personal letter, HARRY E. LeGRAND hydrogeologist U.S.
        Geological Survey, Raleigh, North Carolina, Oct.
        10, 1974.

43.   FENGE. T. "Geomorphological Aspects of the Land Disposal of
        Municipal Solid Waste, " M.A. Thesis, Geography
        Department, University of Victoria, (forthcoming).

44.   McLAREN, J.F. Solid Waste Disposal Study: Stage One.
        Victoria : James F. McLaren Ltd., report submitted
        to the Capital Region District , August 30, 1971,
        p. 3-23.

45.   Ibid., pp. 3-29.

46.   Ibid., pp. 3-29.

47.   Ibid., pp. 3-29.

48.   B.H. LEVELTON and ASSOC. Evaluation of Groundwater Pollu-
        tion at the Hartland Road Sanitary Landfill.Victoria :
        Levelton and Assoc. Ltd. . Report No. 1, Submitted
        to Victoria Disposal Ltd., September 1971, p.3.

49.   FORWARD, C.N. Land Use of the Victoria Area, B.C. Geo-
        graphical Paper No. 43, Geographical Branch, Dept.
        of Energy, Mines and Resources, Ottawa, 1969.

50.   CHILTON, R.R. Climatic Summary for Greater Victoria Region,
        in "An Inventory of Land Resources and Resource Poten-
        tials, " report submitted to the Capital Regional District,
        1973.

51.   LEVELTON, B.H. op. cit., p.4.

52.   As recommended by B.H. LEVELTON and ASSOC. "Evaluation
        of Groundwater Pollution at the Hartland Road
        Sanitary Landfill. Victoria: Levelton and Assoc. Ltd..
        Report No.2, submitted to Victoria Disposal Ltd.,
        June 1972.

53.   This is discussed by HALKETT, I.P.B. The Preservation of Open
        Space on the Saanich Peninsula, B.C., M.A. Thesis.

Geography Department, University of Victoria, 1971.

54. DAWSON, G.M. Report on the Reconnaissance of Leech River and Vicinity. Geol. Survey Can., Report of Progress, 1876-1877, pp. 95-102.

55. HALSTEAD, E.C. Hydrogeology of the Coastal Lowland Nanaimo to Victoria, Vancouver Island, Including the Gulf Islands, B.C. Vancouver : Inland Waters Branch, Department of Energy,Mines and Resources, unpublished manuscript, 1967, p. 37.

56. Ibid., p.6.

57. Ibid., p. 34.

58. HALSTEAD, E.C. "The Cowichan Ice Tongue, Vancouver Island." Canadian Journal of Earth Sciences, 5, (1968), pp. 1409-1415.

59. See Water Table Hydrographs for Wells WR-12-66, WR-2-66, and WR-102-71 in Groundwater Observation Wells of B.C., Groundwater Division, Water Investigations Branch, Water Resources Service, Department of Lands, Forests and Water Resources, Victoria, B.C., 1974, pp. 12-13.

60. WIKEN, E.B. Landscape Parameters and Interpretation, in "An Inventory of Land Resources and Resource Potentials," report submitted to the Capital Regional District, 1973, pp. 91-119.

61. DUDNIK, E.E. SYMAP - User's Reference Manual for Synagraphic Computer Mapping, Version 5, 1969.

62. GUTHRIE, D.L. "Contour Mapping - An Essential Tool Revisited," J.A.W.W.A., 67, No. 7, (July 1975), pp. 389-390.

63. FENGE, T. op. cit.

64. LeGRAND, H.E. op. cit. , p. 963.

PLATE 6
Car crushing, Hartland Road Sanitary Landfill.

CHAPTER 8

SEWAGE DISPOSAL TO THE SEA [1]

Derek V. Ellis

University of Victoria

INTRODUCTION

Disposal of sewage to the sea, without treatment, has always been attractive to the authorities of the Victoria region, and remains so to this day. It is still supported on the traditional grounds that it achieves public health goals while exerting a minimum financial burden on the taxpayers.[2] Examples of this long-standing approach to the sewage problem include the Clover Point sewer (Figure 1,8) built in 1894 to serve the needs of the now 'downtown' area and which still functions; and the Macaulay Point outfall, a new trunk sewer brought into operation as recently as 1972.

Such recent construction as this latest sewer in the series may give an impression of indifference on the part of the authorities[3] to changing public attitudes, as expressed, for example, by demands for "clean" water for recreational use, and by objections to the marine disposal of nutrient-rich human waste which is potentially recyclable to land as fertilizer. Such public attitudes have been apparent in the Victoria area[4] for some time and despite the construction of Macaulay Point outfall it is not justifiable to conclude that the authorities are indifferent to them.

In part, the degree to which authority is responding to such changing public attitudes is shown by the design specifications. In

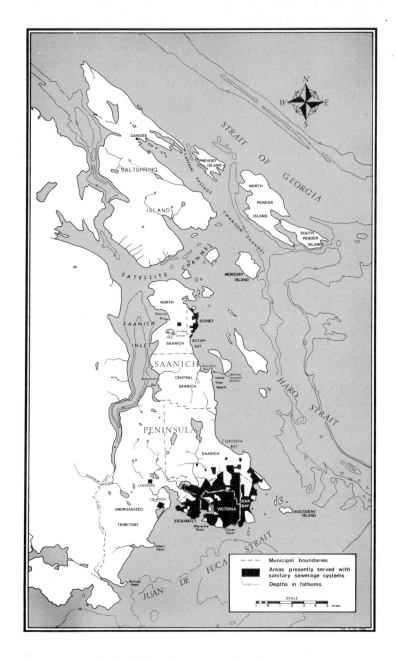

FIGURE 1,8   Location and major sewers.

1894, the beach-discharging Clover Point sewer (then well removed from the downtown area) was considered acceptable. In contrast, the 1972 Macaulay Point outfall is 6000 feet long and 200 feet deep, disposing sewage far offshore. In addition and more significantly a policy of rational decision-making as to the degree of sewage treatment[5] needed has also been implemented. Decisions are now being based on estimates of sewage dilution, and to some extent on the assimilative capacity of the receiving waters. As a result there are now two small sewers (Sidney and Central Saanich) in the Greater Victoria area discharging secondarily treated effluent. Several others are planned.

It should be noted, however, that in the Victoria area, despite treatment, the end result is the same; the sewage is eventually discharged to the sea[6] and no component (such as sludge) is disposed to land. In two cases a relatively "clean" effluent is discharged from man-made treatment plants while the remaining sewers discharge "dirty" effluent and the treatment system is solely the biological assimilative capacity of the receiving sea water.

The three main objectives of this chapter are to review the marine environmental aspects of the Greater Victoria area's sewerage systems, to assess the success of the system in avoiding environmental problems, and to indicate trends in environmentally-based sewage-disposal decision-making in the area. It is not the intent here to examine the potential for alternative sewage disposal systems, for example recycling to land, although the importance of this topic is fully recognized.

The third objective noted above, that of indicating decision-making trends, largely determined the geographical area discussed. This extends from William Head (Figure 1,8) in the south (a federal government reserve within unorganized territory) to Ganges in the north. Included then are the urban waterfronts of the municipalities of Esquimalt,

Victoria, Oak Bay, Saanich, Central and North Saanich and Sidney.
William Head was chosen as the southern extremity of the area under
consideration because disposal here has recently been the subject of
"good practice", and as such is a case-history of significance. Similarly
the northern extremity was located at Ganges because recent regulatory
agency decisions concerning disposal by that settlement appear to have
set precedents which may have considerable future impact.

THE EXISTING CONDITIONS

A comprehensive review of then current sewage disposal conditions and future requirements was undertaken for the region in 1966,
and is the basic reference work for the area.[7] Some of the proposals, in
this report, with or without subsequent modification, have since been
implemented. As a result a new list of existing outfalls is needed.

Furthermore, since 1966 the Provincial Pollution Control Branch
has begun to insist upon substantially better environmental data gathering
and the preparation of detailed impact statements. The data obtained
as a result of these requirements, and information gathered by a decade
of continuing basic oceanographic studies, together allow a better understanding of the local marine environment and its reaction to sewage discharge.[8] Local variations are also more fully appreciated.

The Sewer System and Outfalls

Community outfalls, either existing or planned, in the Greater
Victoria area and Saanich Peninsula (with the exception of those on
Federal Reserves), are shown in Figures 1,8 and 2,8. Documentation
of design and flow characteristics is given in Table 1,8. Discharge
volumes range from 75,000 to 12,000,000 gallons per day. This volume
of sewage can better be visualized by comparison with the 3,000,000 -
10,000,000 gallons per day of controlled flow released down Victoria's

292

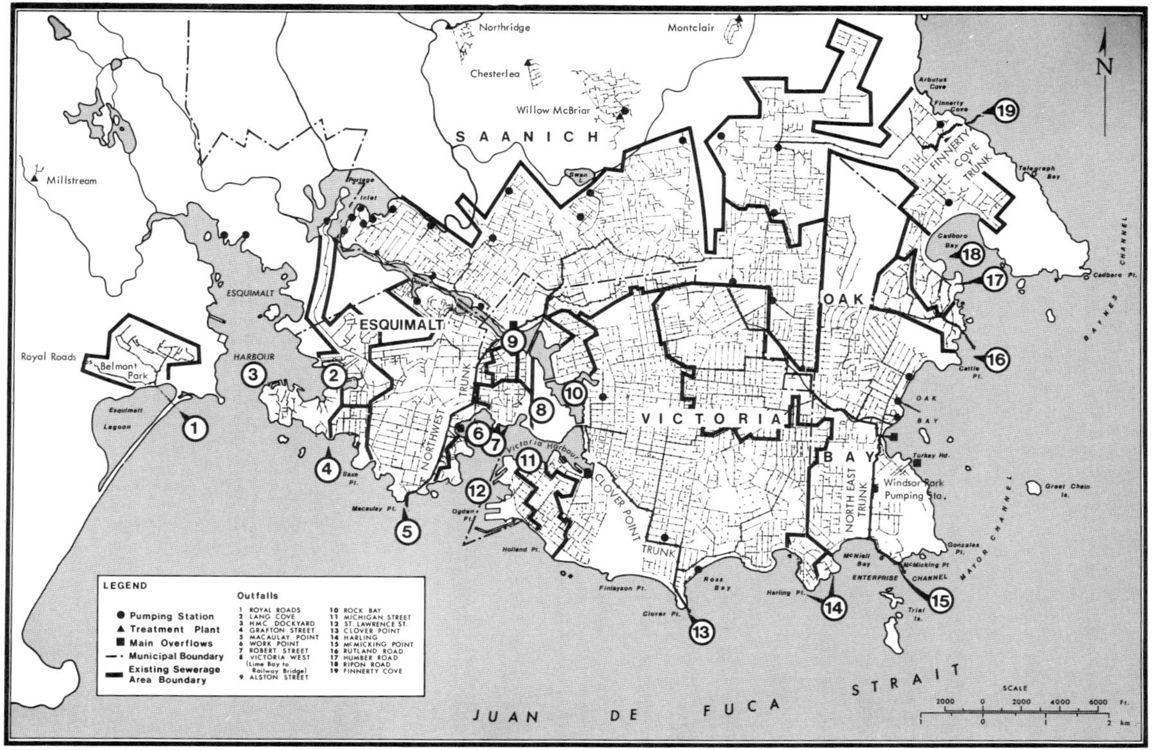

FIGURE 2,8   Sewers of Greater Victoria.

nearest salmon river, the Goldstream.[9] For readers unfamiliar with this waterway, it is noted that a fully open bathtap releases 20,000 gallons of water per day.

Sewage quality ranges from completely untreated effluent, to that produced by secondary treatment, in which there is a reduction of pollution-creating suspended solids, oxygen demand, smell and pathogens. Such treatment, however, appears to have little effect on the quantity of contained nutrients, and it is significant that the accepted level of effluent quality monitoring is such that the nutrient content of most of the discharged sewage cannot be authoritatively documented. The level of sewage treatment would also have an effect on contained industrial contaminants, but this appears not to be of major importance, in the Victoria area, with its small industrial base. There are however, regulations imposed by the Capital Regional District on the quality of wastes admitted to the trunk sewers.[10] These ensure that the onus falls on industrial plants to dispose of wastes which would otherwise affect the District's ability to meet its Discharge Permit restraints.

Effluent quality and discharge points associated with recently installed treatment systems, in the area, are essentially compatible with objectives expected to be accepted by the provincial regulatory authority, in the near future.[11] The intent of these objectives is to ensure that a combination of effluent quality and dilution capacity of the receiving area will be sufficient to reduce fecal coliform levels, on the nearest shoreline, below 200/100 mls.[12] Such a shoreline water quality appears to ensure infection-free conditions in cold temperate regions which are relatively free of enteric diseases. This standard also provides a very considerable safety factor. As a result, recreational use of beaches can be permitted. To place these objectives in the perspective, it should be realized that such a coliform level could be approximated by stirring up one teaspoon of human faeces in a normally filled bath-tub. Such

MAJOR PRE-1966 OUTFALLS (IN ORDER OF DECREASING DIAMETER) AND
POST-1966 ADDITIONS TO THE VICTORIA AREA SYSTEM

Designation location	Sewage type discharged	Outfall specifications			Year of construction	Year of close down**
		Diameter (inches)	Depth in (feet)	Distance offshore (feet)		
Pre-1966						
Michigan Street	Septic Tank	42*	5	500	1965	
Macaulay Point	Raw	36	3	0	1913	1971
Butland Road	Raw	24*	18	730	1917	
Humber Road	Raw	24*	26	375	1917	
Finnerty Cove	Comminution & Chlorination	24	44	1,300	1961	
Clover Point	Comminution during summer	22	3	0	1894	
Ocean Avenue (Sidney)	Raw	21*	35	700	1966	
Lang Cove	Septic Tank	15	0	0	1943	1972
Grafton Street	Raw	14	0	0	1913	1972
Patricia Bay	Comminution, Chlorination available but not used	14	3	350	1943	
McMicking Point	Raw	–	3	0	1913	
		or Design Flow (gpd)			Proposed start–up	Actual start–up
Post-1966						
Macaulay Point	Comminution & Multiport diffuser	12,000,000		6,100	1971	1971
Clover Point	Comminution & Coarse screening	9,350,000			1972	
Sidney	Activated sludge, Secondary treatment	500,000		1,000	1973	1973
Central Saanich	Extended aeration Secondary treatment	300,000		1,480	1973	1973
North Saanich	Comminution, biological treatment, Chlorination	120,000		1,000	1973	
Ganges	Comminution, biological treatment, Chlorination	75,000		1,400	1972	Permit issued revoked and appealed

*   Combined with Storm Sewer

**   It should be noted that many small pollutional sources such as septic tank overflows have been eliminated by the trunk sewer
construction programme of the past decade.

water, while repulsive to many, would not appear to be infective.

It is to be noted that several of the sewerage systems, particularly Macaulay Point and Finnerty Cove, have operated to date below existing pump capacity, and flows have not been continuous but intermittent. Such intermittent flow patterns influence dilution behaviour of a discharge effluent field, and complicate environmental assessment.

Hydrography and Oceanography

The basic hydrography and oceanography of the area was outlined in the 1966 report[13] and in a later review.[14] These detail the broadscale patterns extending through the Straits of Juan de Fuca and Georgia. Since 1966 there has been considerable additional data gathering research by many government agencies and university departments. Much of this information is referenced by the Capital Regional District[15], but some of the work is still in progress and has not yet been documented.

The southern part of the area, between William Head and Discovery Island (Figure 1,8) has been investigated by research teams from the University of Victoria, consulting engineers, federal and provincial agencies. The eastern section, stretching from Discovery Island to Sidney has been the subject of less investigation, although there have been studies at the Finnerty Cove outfall,[16] Island View Beach[17] and Saanichton Bay.[18] Ganges is a peculiar case which will be described in detail elsewhere.

The marine waters north and west of Saanich Peninsula have been very intensively studied. This is because the environmental peculiarities of Saanich Inlet (which is a fjord with a stagnant trough) make it a very suitable site for field experimental studies of food chains and controlled pollution.[19]

The southern marine environment, off Victoria, is a broad shelf, approximately 100 meters deep, extending out to the trough of the Strait of Juan de Fuca. There is a similar, but narrower, shelf extending along the east coast of Saanich Peninsula. This shelf gives way to a more steeply sloping sea bed around the remainder of the Peninsula which then grades into a 200 meter 'pit' in Saanich Inlet.

Water quality is dominated by two current drifts; the first, a surface flow from the north comprising Fraser River discharge, mixing as it passes south; and the second, a deep inflow entering the Strait of Juan de Fuca and upwelling at its head. The intermixing between these two predominant flows is dynamic, responding to irregular tide and wind patterns,[20] and seasonal river discharges. Consequently, salinity, temperature, and current velocities and directions display fluctuating surface characteristics within fairly broad ranges. Salinities, away from shore, may vary between 29 and 32°/oo, and temperatures through 5 to 15°C. There are noticeable annual cycles, tide cycles, irregular wind and river effects (including coastal-stream winter-storm run-off, and summer Fraser River melt-water freshet influences). Local embayments may generate persistent or fluctuating eddy or estuarine conditions, with potential for localized ecosystem disturbances due to stagnation or enrichment. In general, however, tidal currents and the two persistent drifts (surface and deep) generate good dilution capacity away from the shore line. With a few very local exceptions, nutrient deficiency does not appear to be limiting to biological growth, and thus nutrient additions do not normally cause the problems associated with eutrophication (such as phyto-plankton blooms, sedimentation) except in such extreme embayments as Portage Inlet[21] (Figure 2,8).

Pronounced stratification of the water column rarely occurs even in summer. It is only found to take place when exceptionally persistent summer insolation is experienced or intrusion of relatively unmixed Fraser River water occurs or both take place concurrently. Lack of such stratification ensures that a sewage plume, from a deep outfall, will float to the surface and will not layer at a sub-surface density level at which it will disperse horizontally. Consequently, outfall depth-distance combinations have to be great enough to meet surface dilution objectives by vertical mixing alone, without assistance from horizontal sub-surface dispersion.

CASE - HISTORIES

Macaulay Point - A Long, Deep Outfall

In 1970 the Capital Regional District was granted Discharge Permit No. 270-P, permitting replacement of the Macaulay Point beach sewer with a long (6000 feet) deep (200 feet) outfall. Receiving area restraints were:

> The median M. P. N. coliforms per 100 ml shall not exceed 1,000 during the period May to September, inclusive, such samples to be taken approximately 1 foot below the surface in an area where the water depth is between one and three feet. Furthermore, there must be no waste or material attributable to sewer outfalls visible on the beaches or in the waters adjoining the beaches, at any time of the year.

A monitoring programme was also to be implemented[22,23]. Fifteen months of pre-discharge data were obtained, prior to switch-over in August 1971.

After almost 60 years of continuous beach discharge, identifiable environmental effects were high coliform levels (interpreted as a health hazard requiring beach closure) and abnormally high levels of turbidity, colour, nitrite and phosphate. Slight changes in the shoreline ecosystem, consisting of reduced numbers and diversity of large fixed brown algae (kelps) and a reduction of about 1 foot in the extent to which character-istic shoreline plants extended up towards high water mark were ident-ified. An obvious aesthetic problem also existed, the shoreline being typically dirty and smelly. However, wastes on the beaches were not examined in detail by the programme, because of the difficulty of measuring such parameters objectively.

After discharge was switched to the deep offshore outfall, the swimming restriction could be lifted by local health authorities within a few weeks[24] because of a marked reduction in coliform levels. Other water quality parameters also returned to levels similar to those elsewhere along the shoreline, and by the following summer growth season, there was a return to normal shoreline plant populations.

A further 12 months' post-discharge monitoring established some effects of the discharged sewage at the offshore point. Immediately adjacent to the outfall, but not extending to the monitoring stations a mile distant, sediments were accumulating coliform bacteria, and their constituent fauna was changing to include a slightly different array of species. One of these, a large scavenging hermit crab Paguristes turgidus perhaps reflects the nature of changed food supplies for the benthic ecosystem. There was a statistically detectable change in coliform levels in the receiving water with mean values rising from 167 to 484 MPN/100 mls over a half mile radius. Small oil and grease particles often floated to the surface where they could be seen by the occasional observant boat-operator and the hungry sea-gulls (which aggregated to feed at the plume). Temperature and salinity data suggested the sewage plume frequently surfaced, although this was not usually apparent without intensive searching by an observer. However, the permit restraints on coliform levels and shoreline drift were not infringed.

Since 1972, monitoring has been continued routinely by the Capital Regional District through its own staff. Annual reviews are to be provided to the Pollution Control Branch, although the single volume available to date has simply provided the data,[25] without a review of its environmental significance as required by the permit.

299

Clover Point - Beach Discharge Since 1894

The Clover Point sewer is the oldest still functioning in the study area, although plans for its replacement, by a deep outfall are currently being prepared. In 1971, after 77 years of operation, a detailed environmental assessment study was undertaken on behalf of the Pollution Control Branch.[26] As with many of the local studies, new techniques had to be developed (Figure 3,8).

The sewage field was easily identifiable by various traditional, and new, scientific techniques used to measure its dilution and spread. It could, for example, be detected to distances of 2,000 - 5,000 feet down - current from the outfall, but generally stayed at the surface to about 10 feet depth and did not always impinge onshore. Although there was some depression of shoreline algae beds around the outfall, little apparent sub-tidal effect (other than considerable discharged garbage to depths of 50 feet some 200 - 400 feet offshore) could be found. Chlorophyll crops tended to be high, but control stations at other nearby peninsulas also gave high values. However, the ratio between carbon uptake and chlorophyll crop, both in the sewage plume and under laboratory conditions, indicated a depressing effect of the sewage on primary biological production, at least within the visible plume. The results are similar to those obtained around the old Macaulay Point outfall and it is to be expected that there will be a substantial and rapid clean-up, at the time the beach outfall is replaced by a longer outfall.

Attempts to improve the situation at Clover Point were given momentum in 1971, at which time it appeared that a permit for modification of the existing system would only be granted if an appropriate land allowance was made for the construction of a treatment system if needed. Consequently, the procedure recommended by the design engineers was a trunk connection to the Macaulay Point outfall, the land base for which met the

FIGURE 3,8
Clover Point sewer at low tide.

treatment plant space requirement. In 1973 a long outfall discharging
directly at depth to the sea was permitted by the Pollution Control Branch
but, "suitable land has been reserved at Macaulay Point and when treatment becomes necessary the Clover Point and Macaulay Point influent
trunks will be diverted to the new treatment facility and existing outfalls
used for storm water discharges".[27]

Finnerty Cove - Off the Beach

Just south of Finnerty Cove (Figures 1,8 and 2,8) a 1,300 foot
outfall was built in 1961 to discharge the chlorinated and comminuted
wastes of eastern Saanich Municipality to the sea, at a depth of 44 feet.
In 1966, an assessment[28] of its effects indicated that it caused no public
health or aesthetic problems. In 1971 another assessment[29] was unable
to detect biological changes along the shore, but relatively high concentrations of chlorophyll-a were found in the area, and diver observations
of large ground fish concentrations suggested that some enrichment might
be occurring. Furthermore, water coliform levels were above ambient
in the immediate vicinity of the discharge point, although this effect
extended only for 100 - 3000 feet, varying apparently randomly, but
possibly with unrecorded fluctuations in discharge rate. The effluent
could be tracked visually by turbidity, and instrumentally, at surface,
dispersing down-current. On one occasion it was tracked for 3000 feet
from the discharge point. With a combination of onshore winds, slack
water and high discharge rate, it would appear possible that the field could
arrive on shore on some occasions. Such occurrences must be rare events,
since there was no build-up of coliform levels or waste matter along the
shore. However, informed local observers were concerned over the occasional
arrival onshore of waste possibly derived from the outfall. They were also
frequently reminded of the presence of the discharge point by flocks of gulls
congregating over and on the water, where the plume surfaced. Water

quality tests established virtually continuous mixing, and no stratification because of strong tidal currents. These ensure that the plume will almost invariably surface, rather than dispersing at an intermediate depth under stabilized surface layers (Figures 4,8 and 5,8).

The 1971 marine studies of the impact of Finnerty Cove outfall were commissioned as part of an exploration by Capital Regional District in recognition of the need for improving sewage disposal in this area because of rapid population growth. As a result of these studies an out-fall extension, but no additional treatment was recommended. This advice reflected the results of the environmental data, and particularly the water quality information. The suggested extension would place the discharge point in deeper water, beyond the effect of the apparent eddy system which may presently recycle the diluted effluent. As yet, such an extension has not been implemented (Figure 6,8).

The South-Facing Embayments:
Victoria Harbour-Portage Inlet and Esquimalt Harbour

Victoria Harbour, the Gorge Waterway and Portage Inlet together form an estuarine embayment which provides the western boundary of much of the city of Victoria. A number of sewers empty into the harbour and the upper regions of Portage Inlet are surrounded by houses, some of which are still serviced by septic tanks. Public concern over the quality of the inlet was expressed as long ago as 1905, at which time no action was taken. A review of the available data base to 1972 is available.[30]

In 1965 and 1966, the embayment was subject to detailed water quality investigations by a Federal Government team, as a result of which its peculiar oceanographic status was described and the level of pollution stated. The inlet is poorly flushed, although it has some fresh-water run-off in winter. In summer there is a reverse estuary condition, with some sea-water inflow and concentration by evaporation. Thus the inlet is a

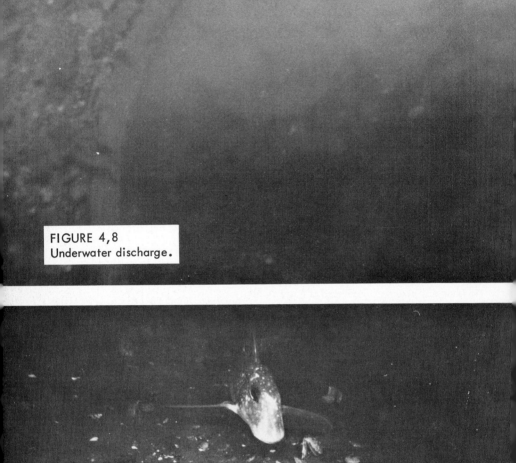

FIGURE 4,8
Underwater discharge.

FIGURE 5,8
Cruising fish near outfall.

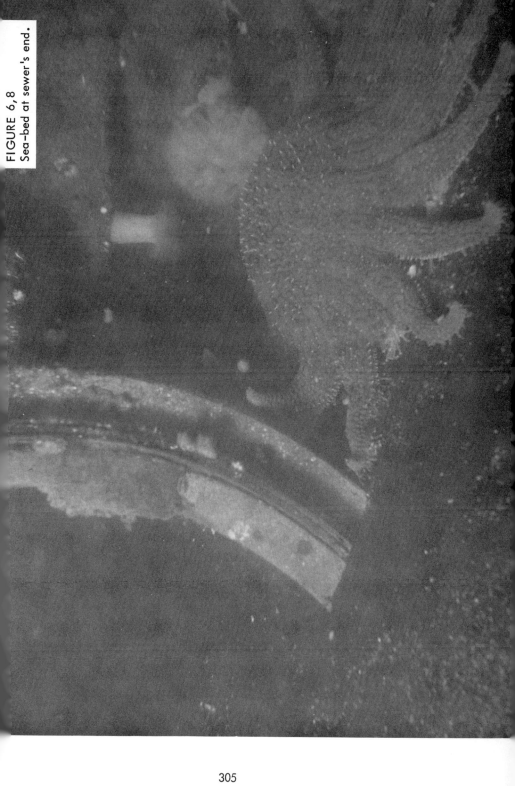

FIGURE 6,8
Sea-bed at sewer's end.

305

sink for pollution.

In the summer the water warms up, increases in salinity while massive nutrient inputs, presumably derived from septic tank seepage (directly into the inlet and its watershed) and from the use of agricultural fertilizers upstream, stimulate eutrophication. There are large scale green algal growths during both spring and summer.

High coliform levels which have led to closure of the beaches for swimming, and recent and on-going siltation are further evidence of pollution. It is to be noted that, of the biological stocks, both herring and water-fowl appear to be maintaining themselves, presumably due to their winter occupation. Herring are dependent on plant growth for egg-laying, and such plants benefit to some extent from the state of enrichment. However, other biological stocks, notably oysters, have declined. It is stressed that this biological information is still largely anecdotal, since only oysters have been the subject of significant scientific investigation. In contrast to the residential and agriculture pollutants of Portage Inlet, Victoria Harbour also has some industrial contamination from heavy metals, and presents the generally dull shoreline facies of a waterfront city.

Esquimalt Harbour is a deep marine embayment, with only limited stream input. Like Victoria Harbour, it receives the discharges from several minor sanitary sewers, but it also has a specialized industrial water-front of federal maritime establishments. At least one study of water quality, related to pollution has been undertaken, but the report has not been made public, since this is a military area. Reference to hydrographic charts suggest that Esquimalt Harbour is a typical estuary with fresh-water run-off at least in winter. It can be assumed that there is probably localized sanitary contamination around the beach sewers which are discharging untreated waste, and in addition some pollution from industrial materials. Occasionally the harbour appears to generate phytoplankton blooms which may be flushed out from this embayment.[31]

Saanich Peninsula - Small Treatment Plants

The Saanich Peninsula is of considerable social interest since it is an area where several concepts of satisfactory sewage disposal have been developed. This region is rapidly becoming urbanized. Population densities and the associated problems of sewage collection and disposal have therefore increased. The end result of this process has been the construction of two secondary treatment plants with deep outfalls which discharge treated effluent, including excess solids, to the sea. Considerable expenditure has been made to minimize discharge of effluent into the embayed and poorly flushed waters of Saanich Inlet, such as Brentwood Bay (to the west of the peninsula) by pumping sewage to the east coast. Improvements are also planned on the federal government reserve at Pat Bay together with the construction of a large oceanographic institute in 1975. The existing 3 foot deep outfall will be replaced by connection to the Sidney trunk sewer system and treatment plant.

The Saanich Peninsula is also of interest in terms of pollution assessment since several of the oceanographic techniques for such assessment have been developed along its coast. A series of investigations were initiated by a joint Federal, Provincial, District team of scientists and engineers in 1964, using the then - novel approach of synoptic current studies utilizing large floats, the movements of which were recorded by transit and aerial photography.[32] On the basis of such evidence, sewage disposal prediction were made, and it was demonstrated that Cordova Bay would be subject to shoreline contamination from certain outfall designs. As a result recommendations on outfall location and design were made.

The discharge sites for small secondary treatment plants at Sidney, Island View Beach and a proposed plant at Bazan Bay have also been the subject of oceanographic studies. Methods used have ranged from traditional drift card records of current flows, to baseline oceanographic and biological data-gathering, appropriate for small discharges.[33]

An innovative attempt to document reclamation, following long-standing sanitary pollution, was initiated in Brentwood Bay. Here biological and water-quality sampling took place prior to the diversion of sewage, from the region to the Central Saanich outfall on the east coast.[34] Additional studies of Brentwood Bay, following sewage diversion, do not appear to have been undertaken. This is unfortunate, since reclamation appears to be rapid, for example some parameters had improved within a few weeks at Macaulay Point after the sewer extension. The public needs assurance that such reclamation does indeed occur.

One further development has taken place in the area. In March 1975, a radar-photographic technique for continuous synoptic broad-scale current studies, developed at the University of British Columbia, for reasons unconnected with sewage disposal[35] was test-operated at Saanichton Bay. The availability of such a technique, however, would appear to have far-reaching future implications for discharge assessment. Several agencies are believed to be installing such appropriate instrumentation.

William Head - Federal Government "Good Practice"

The William Head Penitentiary, a Federal Government institution, implemented design studies in 1974 leading towards construction of a new sewage disposal system. The manner in which these design studies were implemented illustrates recent approaches to impact assessment. The pre-existing, seven, small beach sewers were presumably no longer acceptable to the Federal Government, in its role of " . . . national leadership in environmental protection with respect to its own activities".[36] In consequence, there was to be some form of treatment developed, with possible marine disposal of effluent.

Initial environmental studies, in January and February 1974, dem-

308

onstrated that coliform bacteria levels were high enough to be of concern
at three of ten shoreline stations, but indicated no problem existing at
the proposed discharge point about 100 feet offshore. Shellfish (mussel)
coliform levels, at four of the shoreline stations, were also high. Indeed,
levels were sufficiently high that under the Fisheries Act the mussel beds
could have been declared "Contaminated". After many years of partial
sewage treatment and the marine disposal of effluent, shoreline effects
were determined only microbiologically, although occasionally aesthetic
problems might be apparent during extreme low summer tides.

The authorized environmental baseline studies of the William Head
area, included an assessment of both summer and winter conditions (which
is of course, good practice), but also an innovative attempt to relate
local data gathering with the far more extensive results previously derived
from the Macaulay Point research programme carried out 7 miles to the
northeast. Unexpectedly, it had to be concluded that there were sufficient environmental differences between the two areas to preclude using
data from one to support inferences about the other. This is one of several
sets of data, now becoming available, which indicate that the orthodox
concept of control stations for assessing the results in test areas can be
difficult to apply in the dynamic marine environment.

Ganges – Public Controversy

Ganges, a small town on Saltspring Island, has a serious sewage
disposal problem. Low, marshy ground, within the settlement, receives
contaminated drainage which eventually is discharged into the sea, thus
precluding swimming in a nearby bay and shellfish gathering. It has also
been found impossible to satisfactorily dispose of hospital wastes into the
ground and these contaminate a creek which discharges into Ganges Harbour.

An attempt to develop a better sewage disposal system was ini-

309

tiated in 1965.[37] The resulting proposal, which involved secondary treatment and discharge of effluent, by submerged outfall, progressed through various planning stages until an application for a Discharge Permit was made in December 1972. Alternatives to such marine discharge were considered and rejected on geographical and financial grounds. Three months prior to the permit application, the author of this chapter was asked to undertake appropriate baseline environmental studies, giving due consideration to the restricted tax-base for the area. A small study of appropriate parameters was implemented in January 1973, after the system design had been finalized and the permit application made. The pace of change in environmental matters is now so rapid, that care should be taken to evaluate the environmental assessment procedures adopted, at Ganges, against the standards implied by conventional practices of the late 1960's and early 1970's. At that time, the introduction of marine biological surveys, even late in the design process, was not standard practice by engineers and dischargers and must be regarded as innovative.

Ganges was granted a Discharge Permit, not routinely but only after a public hearing. A province-wide environmental society (SPEC) acted to assist a local objector, and presented a reasoned brief through a staff biologist. Although the permit was granted, with a substantial monitoring requirement, an appeal to the Pollution Control Board was lodged by the objector. A further public hearing followed, again with a reasoned submission by the objector. Eventually the permit was revoked on the grounds that "the applicant has failed to show that the works proposed will not cause pollution."

In the light of recent developments the Ganges effluent disposal problem is recognizable as a particularly complex one, involving geographic, social and biological difficulties. There has been a recent rapid increase in informed public interest in effluent disposal, together with an expansion in the quality of information, considered necessary, before developments (potentially damaging to the environment) are allowed to proceed. Added to this is a cost constraint, since it is perhaps unwise to spend substantial amounts of tax payer's money to obtain design information when risk still remains that the public development planned may not be allowed to proceed.

In reviewing the Ganges case, it seems that the embayed nature of the proposed outfall discharge point aroused concern amongst local citizens about conventional engineering design. As a result, the public have taken responsible action to object to the proposed solution through established channels. The apparent result, however, is that by mid-1975 after 10 years planning, Ganges must continue to tolerate its present sewage disposal system. No resolution to its associated problems appears in sight.

The Ganges case - history is particularly revealing in that this area provided a forum where changing public attitudes and those of environmental groups, government development and regulatory bodies, consulting engineers and environmental specialists could be debated. It should also be noted that an appeal agency has accepted the objection of laymen even though a variety of experienced civil engineers at the design, implementation and pollution control level found the resolution satisfactory. Unfortunately this outcome has not helped the town resolve its immediate pollution concern, but may lead to an optimum resolution of Ganges problems and those of similar "difficult" areas in the future.

Developments? - Albert Head and McMicking Point

There are other developments underway, or needed, in the Victoria area, which can now be appropriately described. In 1974, the Capital Regional District called for proposals for the design of a sewage disposal system for the Colwood-Langford electoral areas.[38] Such proposals were to include appropriate oceanographic and marine biological surveys, the former to provide data for the design and location of a submarine outfall,

311

and the latter to document pre-discharge conditions. Several of the proposals reflected recently acquired environmental capability within consulting agencies. The studies are now in progress. The terms of reference for participation, precluded disposal systems alternative to marine disposal, but required that the studies be undertaken to determine the degree and type of sewage treatment to be selected, and the plant location to be employed, when treatment became necessary.

Finally, in the south of the Victoria area, there remains one major trunk sewer still discharging untreated sewage into shallow water (3 feet below low water mark). This reaches the sea at McMicking Point in Oak Bay where the outfall is somewhat hidden by rocky ground. There has been no specific assessment of the biological consequences of this discharge, although the general results obtained from the study of Macaulay Point and Clover Point can be expected to apply here. That is local unsanitary conditions and minor biological disturbance are currently occurring which will rapidly clear after removal of the pollutant source. Some relevant physical-chemical environmental studies are available, however, since they were components of broader research programs on the effects of the adjacent Clover Point sewer,[39] some 2 miles to the west. Some of the sampling and recording for these studies has been carried out close to the McMicking Point outfall. It should be noted that the planning, design work and proposed construction of the Clover Point system envisages connection of the McMicking Point sewer drainage area to the Clover Point Trunk, and, therefore, the elimination of the McMicking Point Outfall.

A DECADE OF CHANGING ATTITUDES: 1965-1975

The author selected, somewhat arbitrarily, 1965 as the starting point for the preceding résumé of changes in public attitudes concerning

312

effluent disposal. At that time, the Capital Regional District was considering implementation of an overall sewage disposal plan,[40] and followed conventional procedures, for a coastal area, in opting for marine dispersal. Similarly, the pollution control agency also followed convention in defining restraints in terms of coliform standards and visible material. In 1967, however, a new Pollution Control Act[41] came into force. By 1969 the first of a series of marine monitoring programs in the province was imposed on one of Victoria's new, redesigned outfalls. Six years later, in 1975, demands for the accurate prediction of the environmental consequences of sea disposal have led to the revoking of a previously granted permit, on the grounds that "the applicant has failed to show that the works proposed will not cause pollution."

Studies in many areas lead to the conclusion that the environmental impact of sewage disposal can be assessed. There is a strong argument in favor of the view that the costs of such assessment should be met at the design stage, lest much greater cost of remedying a serious pollution problem must be met at some future date. Several different factors have combined to cause this shift in accepted standards. These are now reviewed in some detail (Figures 7,8 and 8,8).

Assessment Techniques and Procedures

In 1970, the Capital Regional District was able to meet its requirement to monitor the marine receiving area at Macaulay Point by contracting to a university-based team of scientists. At that time there were no appropriately instrumented and staffed marine environmental consulting companies; nor was the regulatory agency prepared to use its qualified personnel to monitor the receiving area. The techniques to be used in this assessment were not specified in any detail by the regulatory agency.

The situation has now greatly changed. Assessment techniques

FIGURE 7,8
Submersible for under water inspection.

FIGURE 8,8
Oceanographic vessel for pollution monitoring.

have, in part, been reduced to a series of numerical objectives, and
there exists the opportunity to recommend procedures for which absolute
numerical objectives cannot be defined, such as benthic species diversity
assessment. Marine environment consulting groups, with capability for
advanced and sophisticated assessment, have been developed both within
existing consulting engineering firms and by associations of professional
biologists and oceanographers. Indeed, some of these companies have
played a part in developing techniques, such as continuous surface monitoring and profiling of physical and chemical parameters (Figure 9,8).[42]

The Pollution Control Branch has also installed a major computer
data processing facility, with the capability of functioning as a province-
wide data bank. Such a central facility will have an influence on future
environmental assessment programs within the province, by assuming the
financial burden of time-consuming data collation by the individual
discharger, thus permitting more actual data-gathering for the funds
available (Figures 10,8; 11,8; 12,8 and 13,8).

Regulatory Agency Requirements

The requirement, in 1969, by the Pollution Control Branch that
discharge from Macaulay Point be monitored, instigated a series of
similar requirements both in the Victoria area and elsewhere. Later, in
1973, that regulatory agency initiated a public inquiry into objectives for
the disposal of municipal and municipal-type wastes.[43] These may follow
four sets of previously established industrial objectives, and provide
limiting values for such parameters as dissolved oxygen, salinity, temperature, suspended solids, floatable solids, aesthetic reductions, and nutrients.
It is, however, also intended to provide a manual of recommended biological assessment procedures, analogous to the agency's manual of recommended chemical and physical procedures.[44] This presumably will follow

315

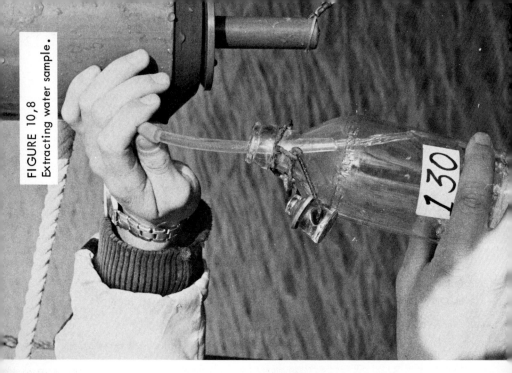

FIGURE 10,8
Extracting water sample.

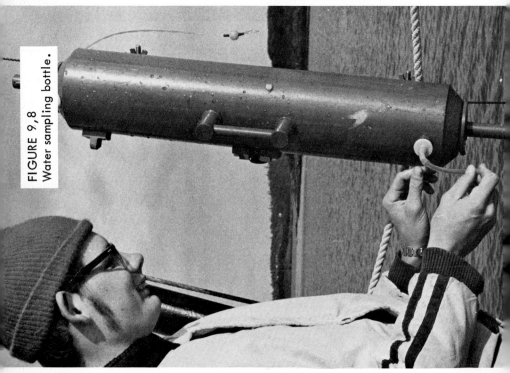

FIGURE 9,8
Water sampling bottle.

FIGURE 12,8
Bathythermography for temperature profiling.

FIGURE 11,8
Secchi disc for water clarity measurements.

FIGURE 13,8
Continuous water sampler.

a format appropriate for the complex problems associated with biological testing.

A further regulatory action, with considerable future significance, was taken in 1975. On appeal, the Pollution Control Board revoked the Ganges Discharge Permit granted by the Director of Pollution Control, some months earlier. The grounds given were that "the applicant has failed to show that the works proposed will not cause pollution." The nature of the Board merits scrutiny, since revocation of a permit after prolonged consideration, by the Pollution Control Branch and its professional staff, is a regulatory action of some significance. In 1967, when the Board's present format was established, it was comprised of representatives of five provincial government bodies (Water Resources, Health, Forests, Agriculture, and Fisheries) and a university M. D. with health specialisation. In 1975, essentially the same government branches were represented, although several of the individuals concerned had changed. A conservationist and a university biologist had also been added to the Board.

"Good Practice" in Design

Draft guidelines[45] for the development of sewage disposal systems on government owned land are revealing about the nature of the concept of "good practice." Attitudes towards environmental problems have changed substantially during the past decade. As late as 1961, the Finnerty Cove sewer could be designed by consulting engineers, accepted by the public's elected representatives and built, with only supporting evidence from bottom profiles obtained by echo-sounder, drift card studies of current patterns, and a few lead-line sediment samples to demonstrate presence of soft bottom or bedrock. By 1969, substantially more environmental information was required before major outfalls could be built. Yet as late as 1972, minor sewers servicing small communities (such as Ganges), required no more environmental information than provided by hydrographic charts. In 1975, permit applications for marine sewers (outside the Victoria area) are still being made which represent an outfall by a line drawn straight out from the coastline on a map devoid of sea-bed depths, contours, and substrate information. The 1975 decision of the Pollution Control Board, on the Ganges permit, may lead towards the recognition that proposals that lack marine information cannot be designated "good practice" by design engineers and the public's representatives. Even if the Board's decision is subsequently revoked, it can still be taken as an indicator of current trends in pollution control requirements.

There is a definite practical problem in requiring such good design practice. This hinges on the extent to which a developer should be required to finance environmental assessment and engineering design details, while it remains possible that permission for the development will not be granted. The necessity for such risky expenditure, an investment which is lost if a permit is not granted, may substantially reduce certain types of otherwise appropriate development.

THE FUTURE

On the basis of the preceding description, certain trends are evident that allow the following predictions of future developments.

Marine Environmental Assessment Procedures

These procedures consist of four components, pre-design studies, pre-discharge studies, design checks and receiving area quality control. Not all of these components are yet equally accepted as "good practice." The future should see increased recognition of all four components, increased effectiveness in utilizing their results, and greater efficiency in

implementation. The extent to which they are employed will be related
to the quantity of sewage discharged and restraints imposed by the funds
available.

Pre-design studies, by appropriate identification of marine environmental hazards and publicly sensitive areas, such as swimming and shellfish beaches, and fishery nursery areas, assist in outfall design and location decisions. In this area there is considerable potential for improved
practice by design engineers, since the information is usually available
in government files and costs for assembling should not be high.

Pre-discharge studies are those designed to provide an assessment
statement of the marine environment and its organisms, as they exist prior
to disposal of sewage. Such studies usually require implementation for
a full year, and should follow the pre-design studies, which have identified the areas of concern. Growing experience of such studies has generated a body of professional expertise able to identify techniques having a
maximum utility.

Design criteria checks are intensive studies which should be
implemented immediately a sewer begins to function. Their purpose is
to demonstrate that the completed system will meet the receiving area
specifications. Such studies are not yet normally implemented. They
should, however, become a requirement, since if any gross infringement
of receiving area specifications, resulting from inadequate design, occurs,
this should be caught early and remedied quickly. In practice, such
problems rarely arise, no examples being available from the Victoria
area. Environmental problems appear more likely to occur from inadequate
sewer maintenance and operation than from inadequate design.

Receiving area quality control consists of routine monitoring of
parameters previously known, or locally determined, to be sensitive to
sanitary sewage. These are essentially attempts to determine when levels
exceed or are trending towards environmentally-disturbing values, so that

remedial action can be taken. This includes altering the quality of the effluent being discharged either by treatment or operational changes. Monitoring of this form can also assist in making rational decisions, during the planning of the extension of existing sewage disposal systems, in areas of population growth.

It is noticable that existing monitoring is undertaken by some dischargers on specific outfalls, such as Macaulay Point, but that the area, south of Victoria currently receiving multiple discharges, can be considered an oceanographic entity and requires synoptic monitoring. Coordinated multiple-discharge receiving area monitoring is required, yet the implementation of such a system has not yet been possible for the Victoria area.

The problem of determining what degree of environmental impact assessment is needed prior to granting approval for a discharge is presently controversial and must be resolved soon, since funds for such assessment are provided by the tax payer and are not unlimited.

Unknowns

There are a number of phenomena about which information is insufficient to make a reliable decision on sewage management.

For several years, nationally-circulated popular media presentations have indicted Victoria for its marine disposal system, basing claims that it is a health hazard on a high infectious hepatitis rate.[46] Infectious hepatitis is a reportable disease and statistics allow comparison of relative Canadian levels. The evidence provided by this data indicates that the infectious hepatitis incidence in the Capital Regional District is no different from that of the rest of the Province of British Columbia.[47] A definitive study is clearly required.

There is a current trend in sewage-treatment literature to condemn

chlorination as a disinfecting procedure because of the production of
carcinogenic chlorinated hydrocarbons that can be experimentally-
induced by this process, and field identification of fish toxicity.[48] Yet
chlorination is still the cheapest and most reliable of the known disinfecting procedures. The provincial regulatory agency has, however,
reacted to this current controversy by recommending dechlorination to
produce very low chlorine residuals, in receiving areas, with dilution
levels between 20:1 and 200:1, or where significant effects may be
demonstrated on fish in lakes. It has, however, taken alternative steps
to remove possible biological hazards from chlorination by recommending
that the use of chlorine be minimized and that chlorine be non-detectable,
in a receiving area, by the most sensitive of the standard testing procedures.
In addition, the initial dilution zone (where chlorine residuals at levels
suspected of marginal toxicity may be dispersed by dilution) must not
infringe on significant biological resource areas such as fish migration
routes or shellfish beds. It appears that better information is needed on
the toxicity of chlorine as a disinfectant, particularly in sea-water, since
there are productive shellfish beds around the Victoria district coastline
and important herring spawning grounds in the north. Until this data
becomes available, chlorination is being eliminated from disposal systems
where not essential, for example, it was not required at the Macaulay
Point deep outfall.

There appears to be one other major unknown in sewage disposal
decision-making in the Victoria area and elsewhere. This is the quantity
of nutrients being discharged, relative to the assimilative capacity of the
receiving area. Such information would indicate when, if ever, enrichment will become a problem for an area, and would thereby assist in
indicating the need for considering alternative disposal systems.

Alternative Disposal Systems

During the recent development of marine disposal systems, pro-

323

ponents of systems alternative to marine discharge, have been active in the Victoria area. There is also considerable pressure from outside the area against marine disposal.

Whether alternative systems will be adopted in the future is still very uncertain. The cost differential, as expressed by conventional engineering analysis, is so great that taxpayers should look dubiously at alternative systems. Nevertheless, conventional engineering analysis is itself subject to evolution, and the escalating rate of change in the concept of "good practice" is the current dominant force in sewage disposal decision-making. It remains to be seen if the present rate of change will be maintained and if so, whether it will lead to sewage disposal system alternatives to marine discharge.

CONCLUSION

Beach and shallow discharge of untreated sewage in the Victoria area is now being replaced by other disposal procedures, which nevertheless still involve the marine disposal of an effluent. Where replacement has been monitored there has been reversion to what can be considered normal environmental conditions within a few weeks or months after shallow water discharge ceased. The deep outfalls, in the Victoria area, discharging sewage with degrees of treatment ranging from comminution to secondary, have in general only subtle effects on the marine environment, which require technical procedures for their detection.

Over the past decade there have been significant changes in the degree of pollution control exercised by all levels of the operational and regulatory authorities. In particular, during the past two years the rate of change appears to be increasing, and such a process must have a considerable impact on future "good practice". However, it is unlikely that the changes will lead to early abandonment of marine disposal.

324

REFERENCES

1. This review compiles information assembled from numerous sources
    since 1970. Many people have co-operated in providing
    assistance either voluntarily, or through their employment
    in responsible administrative positions. Regretfully, I can-
    not acknowledge them all individually, but two people
    must be singled out for my appreciation. One is Mr. W.
    Gerry, P.Eng., Engineering Director, Capital Regional
    District, and the other Mr. E. Dew-Jones, Chief, Mun-
    icipal Division, Pollution Control Branch. Both have been
    helpful in providing information. Several agencies have
    funded the research and literature reviews of which the
    Pollution Control Branch of British Columbia, the Capital
    Regional District of British Columbia, the National Research
    Council of Canada, and the Faculty Research Fund of the
    University of Victoria have borne the major costs.

2. Cost comparisons are provided by the Capital Regional District. Mun-
    icipal Sewage Disposal by Marine Outfalls: Future Planning,
    Discharge Permit Requirements, Environmental Impact and
    Monitoring Techniques. Submission to the Public Inquiry
    into Municipal and Municipal-type Waste Discharges in
    British Columbia 1973. As an example of relative capital
    costs between sea- disposal without treatment and sea-dis-
    posal with treatment, capital cost estimates for replacement
    of the Clover Point beach sewer can be quoted. Sea-dis-
    posal without treatment $5,225,000. Sea- disposal with
    primary treatment $14,900,000.

3. The "authorities" in this context means the complex of government
    agencies responsible for sewage disposal in the Victoria
    region, and for the implementation of regulations for the
    design and operation of sewage disposal systems. In the
    Victoria area, the Capital Regional District acts as a cen-
    tralized planning and operation agency for the various mun-
    icipalities and the unorganized areas. The Capital Regional
    District is one of a series of province-wide planning and
    administrative agencies comprising representatives from the
    component municipalities and unorganized areas. Provin-
    cial government involvement lies through the Director of
    Pollution Control and his agency the Pollution Control
    Branch. The Branch essentially administers the Pollution
    Control Act 1967 subject to appeals to, and policy decis-
    ions by, a Pollution Control Board. Other departments

such as the Health Branch may have powers ranging from
advisory to enforcement in certain cases. Federal government involvement is usually indirect in the sense that it
maintains channels of communication with the Pollution
Control Branch at the time applications for Discharge
Permits are being considered. However, Federal Government involvement can become direct if the discharger in-
fringes pollution control measures such as in the Fisheries
Act 1952, which for example prohibits the "deposit of a
deleterious substance of any type in water frequented by
fish"... 33 (2). The Federal Government also has the
authority to declare an area "Contaminated" which it does
using a variety of criteria, and such a declaration precludes
the taking of shellfish (molluscs) from that area.

4.   As examples, a series of papers in 1969 and 1970 by Dr. M. A. M.
Bell, Biology Department, University of Victoria can be
referenced. These were prepared at the time when application for a Discharge Permit for the new Macaulay Point
outfall was being made. They are "Waste Management and
Environmental Quality in Victoria, B.C." in MAUNDER,
W. J. (ed.) Pollution. University of Victoria, 1969,
"Sewage and Refuse Disposal in Land" Journal of Environmental Health, 32, 2, p. 183-189, "Waste Disposal Alternatives may Dispel Urban Crisis" Idem 32, 4, p. 389-
393, "Interdisciplinary Cooperation needed for Waste
Disposal" Idem 32, 6, pp. 686-691.

5.   Terminology for degrees of treatment will follow this system: "comminution" - grinding, hence reducing the size of solid
material, "primary" - settling of solids in holding tanks,
"secondary" - various techniques to increase biological
decomposition, anti-pathogen activity and settling of
sewage, "tertiary" - advanced techniques producing a
high quality, even drinking quality, effluent. In addition,
the practice of "disinfecting" is commonly applied, usually through "chlorination".

6.   Personal communication W.G. Gerry, Director of Engineering,
Capital Regional District, Victoria, B.C. 1975.

7.   Associated Engineering Services Limited. Sanitary Sewerage Study
of the Greater Victoria Area. Report to the Joint Sewerage Committee, 1966, 237 pp.

8. Environmental principles underlying waste disposal to the sea have been reviewed by WALDICHUK, M., "Waste Dispersal in Relation to the Physical Environment - Oceanographic Aspects." Syesis 1, 1968, pp. 4-25.

9. Personal communication R. Upward, Commissioner, Greater Victoria Water District, April 11, 1975.

10. Capital Regional District. Regulations Governing the Admission of Wastes into Sewers, 28th October 1972.

11. Pollution Control Branch. Pollution Control Objectives for Municipal Type Waste Discharges in British Columbia (forthcoming).

12. Personal communication W.K. Oldham, Associate Professor, Department of Civil Engineering, University of British Columbia, and Member of the Panel of Technical Advisors to the Director of Pollution Control for the Public Enquiry into Municipal Type Discharges, 1973.

13. See reference 7.

14. WALDICHUK, M., op.cit. See reference 8.

15. See reference 2.

16. Ibid.

17. Ibid.

18. Saanichton Bay is under environmental study for other reasons that sewage discharge, but the information adds to the relevant data bank. This particular area also clearly reflects the rapid change in public attitudes to needs for environmental assessment. In early 1974 government recommendations initially favouring a marine development were based on some prior literature plus two small project-centered studies. Within 6 months by late 1974 a comprehensive environmental survey encompassing long-shore drift, eel-grass, fisheries, waterfowl assessment and social responses was initiated as a result of public protests. A rough assessment of the cost of contracted projects and the time of civil service professional and technical employees by early 1975 indicated costs of the order of $100,000. It is the

author's experience that prior to 1975 a consultant's proposals for environmental assessment at costs greater than about $5000 - $10,000 were treated as academic and unrealistic. The development finally could not proceed through refusal of the provincial government to grant a foreshore lease. This is a timely example of the extent of the problem involved in how much should be spent on design details and environmental assessment prior to approval from the regulatory agencies.

19. Literature on Saanich Inlet oceanography and marine ecosystem is extensive, see reference 8. In addition an international controlled pollution experiment (CEPEX) is now in progress in the area, and further information is to be expected in the next few years.

20. Canadian Hydrographic Service. Canadian Tide and Current Tables, and Atmospheric Environment Service. Annual Meteorological Summaries and Monthly Meteorological Summaries. Department of the Environment.

21. A review of environmental data from the Portage Inlet system was collated in 1972. ELLIS, D.V. (ed.) Biology of the Gorge Waterway System, 6 volumes. Report to the Capital Regional District of British Columbia. Significant papers are WALDICHUK, M., "Eutrophication Studies in a Shallow Inlet on Vancouver Island," Journal Water Pollution Control Federation 41, 5 (1967), pp. 745-764. FOSTER, H.D., "Geomorphology and Water Resource Management: Portage Inlet, a Case Study on Vancouver Island," Canadian Geographer, 16, 2 (1972), pp. 128-143. CARNE, R.C. A Study of Trace Metal Content of Victoria Harbour, B.C. Bottom Sediment. B.Sc. Honours Thesis, University of British Columbia, 1974.

22. BALCH, N., ELLIS, D. and LITTLEPAGE, J. Macaulay Point Outfall Monitoring Program, May 1970 - October 1972 Final Report, 2 volumes. Report to the Capital Regional District of British Columbia, 1973. This final report summarizes and reviews data-gathering by the University of Victoria monitoring team. It also references the series of nine data reports and two annual reviews produced by the programme. Subsequent published papers on components of the programme are BALCH, N., BROWN, D., PYM, R., MARLES, E., ELLIS, D. and LITTLEPAGE, J., "Monitoring Marine

Outfalls by Using Ultraviolet Absorbance," Journal Water Pollution Control Federation, 47, 1, pp. 165-202. And BALCH, N., ELLIS, D., LITTLEPAGE, J., MARLES, E. and PYM, R., "Intensive Initial Monitoring of a Deep Marine Sewage Outfall," Journal Water Pollution Control Federation (in Press). Full details of the shoreline effects on plants are given in COON, L.M. Effects of Untreated Sewage Effluent on the Ecology and Metabolism of Intertidal Flora on Rocky Shores Adjacent to Shoreline Discharging Sewage Outfalls. University of Victoria, M.Sc. Thesis, 1973. It should be noted that the coliform restraint at Macaulay Point refers to total not fecal coliforms (for which the equivalent standard is 200/100 mls.).

23. The monitoring programme at the Macaulay Point Outfall was the first of a series of such programmes imposed by the Pollution Control Branch on marine discharges. The letter of transmittal (dated April 28, 1969) for the permit specifically indicated that the programme would serve in a research capacity.

"Although there is a wealth of information throughout the world in support of the proposed method of disposal authorized, we have decided in the public interest to use the proposed outfall as a research project.... It is the intention of the Pollution Control Branch to use the information gathered, in conjunction with additional work to be undertaken by us, to verify the effectiveness of the proposed method of disposal and to document for future reference the suitability of long outfalls for the disposal of sewage in the coastal waters of British Columbia."

In 1971 three further discharges also had monitoring programmes imposed: Permit number 379-P, Utah International (a copper mine and concentrator), Number 395-P Electric Reduction Co. of Canada (a chemical plant), and Number 427-P, Jordan River Mine (a second copper mine and concentrator). The Utah programme was also substantial and will continue to 1976, the Jordan River mine programme terminated in December 1974 when the mine closed. The Jordan River receiving area, however, is now the subject of a reclamation research study contracted by the Pollution Control Branch. Since 1971 there have been other marine monitoring studies imposed on dischargers from time to time.

24. The Capital Regional District received a letter dated October 5, 1971, from the Greater Victoria Metropolitan Board of Health advising that the beaches near the Macaulay Point Outfall were satisfactory for swimming.

25. Capital Regional District, Macaulay Point Outfall Monitoring Program, November 1972 - September 1973, 1974.

26. Clover Point design studies are described in reference 2, however, assessment studies are in ELLIS, D.V. Clover Point 1971. An Investigation of the Accumulated Effects of the Clover Point Sewer on the Local Marine Ecosystem; including the Shoreline, 1971: and ELLIS, D.V., LITTLEPAGE, J., COON, L.M. and DRINNAN, R.W. The Clover Point Investigation. Experimental Analysis of the Effects of Untreated Domestic Sewage on Marine Primary Production, 1972. Both are reports to the Pollution Control Branch.

27. See reference 6.

28. DRINNAN, R.W., ELLIS, D.V. and LITTLEPAGE, J.L. Finnerty Cove Outfall: Report on Marine Biology and Water Quality, 1971. Report to the Capital Regional District of British Columbia. There is reference in Appendix 2 of this report to a 1966 study of public health and aesthetic problems. The biological report was part of a larger study recommending the form of additions to the existing system: Associated Engineering Services Limited. Finnerty Cove Outfall Study. Report to the Capital Regional District, 1971.

29. Ibid.

30. See reference 21.

31. Chlorophyll - assessments indicating phytoplankton concentrations in Esquimalt Harbour are described in the reports of the Clover Point outfall survey. See reference 26.

32. KEENAN, C.J., KINNEAR, A.C., SAUNDERS, F.H., TULLY, J.P., WALDICHUK, M., WIGEN, S.O. and YOUNG, R.B. Current Observations in Cordova Bay and Predictions on Sewage Disposal. Fisheries Research Board of Canada, Manuscript Report Series No. 197, 1966.

330

33. The treatment plant outfall location, environmental baseline and monitoring studies comprise a series of small reports to the Capital Regional District by various agencies.

34. Dobrocky Seatech Ltd. Water Quality and Biological Surveys-Brentwood Bay. Data Report for November 1971 - January 1972. Report to the Capital Regional District, 1972. There were two small additions reporting data gathered to May 1972.

35. Personal communication J. Dobrocky, President, Dobrocky Seatech Ltd. Mr. Dobrocky produced his cinerecord in March 1975, based on an instrumentation system developed by Dr. Pond and Mr. Buckley, Institute of Oceanography, University of British Columbia.

36. The William Head design and environmental studies are presently confidential to the Federal Government, but the latter were prepared by the author of this chapter. Federal Government guidelines on wastewater treatment at government institutions have been drafted, but apparently are not yet public documents.

37. The large number of documents on the Ganges problem and possible solutions are summarized to some extent in Capital Regional District, A Brief presenting technical evidence for consideration at the public hearing on 7 March 1974 into the matter of an Application pursuant to the Pollution Control Act, 1967 and dated 21 December 1972 on behalf of the Capital Regional District to discharge treated effluent by marine outfall into Ganges Harbour, Ganges, B.C. 1974. The result of the appeal to the Pollution Control Board was documented in a letter from the Chairman of the Board to the Canadian Scientific Pollution and Environmental Control Society dated March 12, 1975. The transcript of that appeal made available to the author through Pearlman and Lindholm, Barristers and Solicitors, Victoria, is also an important document.

38. See reference 6.

39. See reference 26.

40. See reference 7.

41. Pollution Control Act 1967, Province of British Columbia.

42. DOBROCKY, J.J. and GODDARD, J.M. Continuous measurement
     of physical oceanographic parameters near a sewage out-
     fall. Submission to the Public Inquiry into Municipal-
     type Waste Discharges, 1973. BEAK, T.W., GRIFFING,
     T.C. and APPLEBY, A.G. "Use of artificial samplers to
     assess water pollution," Biological Methods for Assessment
     of Water Quality, 1973, pp. 227-241.

43. See reference 11.

44. Personal communications E. Dew-Jones, Chief, Municipal Division;
     and W. Venables, Director, Pollution Control Branch.

45. The Federal Government is preparing in-house guidelines for planning
     sewage disposal systems at government establishments as
     part of its role in setting standards of "good practice".

46. Personal communication R.O. Brinkhurst, Director, Biological
     Station, St. Andrews, New Brunswick, May 2, 1974.

47. Personal communication Dr. A.S. Arneil, Regional Health Officer,
     Capital Regional District Community Health Service,
     June 3, 1975.

48. PIECUCH, P.J. "The chlorination controversy." Editorial in Journal
     of the Water Pollution Central Federation, 46 12, pp.
     2637. BULL, C.J. and HOOTON, R.S. In Situ Bioassays
     Gold River Municipal Sewage Outfall. In-house report
     by the B.C. Provincial Government and subsequent inter-
     departmental memoranda. BASCH, R.E. and TRUCHAN,
     J.G. Calculated Residual Chlorine Concentrations Safe
     for Fish. Michigan Water Resources Commission, Technical
     Bulletin 72 - 2, 29 pp.

THE CONTRIBUTORS

Charles N. Forward, B.A., M.A., Ph.D., Professor and Head,
  Department of Geography, University of Victoria.
  For a period of eight years before coming to Victoria
  in 1959, Dr. Forward was engaged in research with
  the Geographical Branch, Ottawa. His degrees are
  from the University of British Columbia and Clark
  University.

Michael C.R. Edgell, B.A., Ph.D., Assistant Professor, Department of
  Geography, University of Victoria. Receiving both
  his degrees from the University of Birmingham,
  Dr. Edgell also spent a year attending a post-graduate
  course in conservation at University College London.
  Joining the University of Victoria for the first time
  in 1965, he moved to Australia in 1968 to teach at
  Monash University and returned to Victoria in 1972.

Stanton Tuller, B.A., M.A., Ph.D., Associate Professor, Department
  of Geography, University of Victoria. Dr. Tuller
  received his B.A. degree from the University of
  Oregon in 1966 and both his M.A. and Ph.D. from
  the University of California in Los Angeles, the
  former in 1967 and the latter in 1971. Dr. Tuller's
  research and teaching interests have as their focus the
  applied aspect of micro-climatology.

Rodney R.H. Chilton, B.Sc., Climatologist, British Columbia Environment
  and Land Use Committee Secretariat, Climate and Data
  Services. Mr. Chilton is a 1971 graduate of the
  University of Victoria, where he specialized in
  Geography, with particular emphasis on climatology.
  He is an active member in the Environment 100 and the
  B.C. Naturalists.

Harold D. Foster, B.Sc., Ph.D., Associate Professor, Department of
  Geography, University of Victoria. Dr. Foster
  received both his degrees from the University of
  London, attending University College London from
  1961 to 1967. His initial interest focused on glacial
  geomorphology but more recently he has published
  widely on applied physical geography (hazards and
  renewable energy) and hydrology.

Vilho Wuorinen, C.D., B.A., M.A., received his first degree from Carlton University after accumulating part-time credits from Carleton and the Universities of London, Ottawa, and Victoria. While serving as the Canadian Forces University Liaison Officer at Victoria, he completed his master's degree. After retirement from the Service, he was granted an NRC Scholarship and began work on his doctorate at the University of Victoria, where he is specializing in natural hazard research.

Regan F. Carey, B.Sc. graduated from the University of Victoria in 1975 with a B.Sc. degree in geography. His special interests are applied geomorphology, natural hazards and computerized cartography.

Terence Fenge, B.Sc., received his first degree from University College of Swansea, University of Wales in 1973. He has since been working towards an M.A. degree at the University of Victoria. His major area of research has been the relationships between geomorphology and the land disposal of municipal solid wastes, the subject of his thesis.

Derek V. Ellis, B.Sc., M.Sc., Ph.D., is an Associate Professor, Department of Biology, University of Victoria. His B.Sc. was awarded by the University of Edinburgh, whilst both his M.Sc. and Ph.D. are from McGill. Dr. Ellis is a Marine ecologist who has researched and taught at the University of Victoria since 1964, and who previously worked as a scientist for the Fisheries Research Board of Canada at the Biological Station, Nanaimo, B.C. In 1970 he started a series of marine pollution impact programmes in the Greater Victoria area. This has expanded to involve him in similar studies for government and industry throughout British Columbia, in the Arctic and in Argentina.